T0275588

LONDON MATHEMATICAL SOCIETY LECTURE NOTE SERIES

Managing Editor: Professor J.W.S. Cassels, Department of Pure Mathematics and Mathematical Statistics, University of Cambridge, 16 Mill Lane, Cambridge CB2 1SB, England

The titles below are available from booksellers, or, in case of difficulty, from Cambridge University Press.

34 Representation theory of Lie groups, M.F. ATIYAH *et al*
46 p-adic analysis: a short course on recent work, N. KOBLITZ
50 Commutator calculus and groups of homotopy classes, H.J. BAUES
59 Applicable differential geometry, M. CRAMPIN & F.A.E. PIRANI
66 Several complex variables and complex manifolds II, M.J. FIELD
69 Representation theory, I.M. GELFAND *et al*
76 Spectral theory of linear differential operators and comparison algebras, H.O. CORDES
77 Isolated singular points on complete intersections, E.J.N. LOOIJENGA
83 Homogeneous structures on Riemannian manifolds, F. TRICERRI & L. VANHECKE
86 Topological topics, I.M. JAMES (ed)
87 Surveys in set theory, A.R.D. MATHIAS (ed)
88 FPF ring theory, C. FAITH & S. PAGE
89 An F-space sampler, N.J. KALTON, N.T. PECK & J.W. ROBERTS
90 Polytopes and symmetry, S.A. ROBERTSON
92 Representation of rings over skew fields, A.H. SCHOFIELD
93 Aspects of topology, I.M. JAMES & E.H. KRONHEIMER (eds)
94 Representations of general linear groups, G.D. JAMES
95 Low-dimensional topology 1982, R.A. FENN (ed)
96 Diophantine equations over function fields, R.C. MASON
97 Varieties of constructive mathematics, D.S. BRIDGES & F. RICHMAN
98 Localization in Noetherian rings, A.V. JATEGAONKAR
99 Methods of differential geometry in algebraic topology, M. KAROUBI & C. LERUSTE
100 Stopping time techniques for analysts and probabilists, L. EGGHE
104 Elliptic structures on 3-manifolds, C.B. THOMAS
105 A local spectral theory for closed operators, I. ERDELYI & WANG SHENGWANG
107 Compactification of Siegel moduli schemes, C-L. CHAI
108 Some topics in graph theory, H.P. YAP
109 Diophantine analysis, J. LOXTON & A. VAN DER POORTEN (eds)
110 An introduction to surreal numbers, H. GONSHOR
113 Lectures on the asymptotic theory of ideals, D. REES
114 Lectures on Bochner-Riesz means, K.M. DAVIS & Y-C. CHANG
115 An introduction to independence for analysts, H.G. DALES & W.H. WOODIN
116 Representations of algebras, P.J. WEBB (ed)
118 Skew linear groups, M. SHIRVANI & B. WEHRFRITZ
119 Triangulated categories in the representation theory of finite-dimensional algebras, D. HAPPEL
121 Proceedings of *Groups - St Andrews 1985*, E. ROBERTSON & C. CAMPBELL (eds)
122 Non-classical continuum mechanics, R.J. KNOPS & A.A. LACEY (eds)
125 Commutator theory for congruence modular varieties, R. FREESE & R. MCKENZIE
126 Van der Corput's method of exponential sums, S.W. GRAHAM & G. KOLESNIK
128 Descriptive set theory and the structure of sets of uniqueness, A.S. KECHRIS & A. LOUVEAU
129 The subgroup structure of the finite classical groups, P.B. KLEIDMAN & M.W. LIEBECK
130 Model theory and modules, M. PREST
131 Algebraic, extremal & metric combinatorics, M-M. DEZA, P. FRANKL & I.G. ROSENBERG (eds)
132 Whitehead groups of finite groups, ROBERT OLIVER
133 Linear algebraic monoids, MOHAN S. PUTCHA
134 Number theory and dynamical systems, M. DODSON & J. VICKERS (eds)
135 Operator algebras and applications, 1, D. EVANS & M. TAKESAKI (eds)
136 Operator algebras and applications, 2, D. EVANS & M. TAKESAKI (eds)
137 Analysis at Urbana, I, E. BERKSON, T. PECK, & J. UHL (eds)
138 Analysis at Urbana, II, E. BERKSON, T. PECK, & J. UHL (eds)
139 Advances in homotopy theory, S. SALAMON, B. STEER & W. SUTHERLAND (eds)
140 Geometric aspects of Banach spaces, E.M. PEINADOR & A. RODES (eds)
141 Surveys in combinatorics 1989, J. SIEMONS (ed)
144 Introduction to uniform spaces, I.M. JAMES
145 Homological questions in local algebra, JAN R. STROOKER
146 Cohen-Macaulay modules over Cohen-Macaulay rings, Y. YOSHINO
147 Continuous and discrete modules, S.H. MOHAMED & B.J. MÜLLER
148 Helices and vector bundles, A.N. RUDAKOV *et al*
149 Solitons, nonlinear evolution equations and inverse scattering, M. ABLOWITZ & P. CLARKSON
150 Geometry of low-dimensional manifolds 1, S. DONALDSON & C.B. THOMAS (eds)
151 Geometry of low-dimensional manifolds 2, S. DONALDSON & C.B. THOMAS (eds)
152 Oligomorphic permutation groups, P. CAMERON

153 L-functions and arithmetic, J. COATES & M.J. TAYLOR (eds)
154 Number theory and cryptography, J. LOXTON (ed
155 Classification theories of polarized varieties, TAKAO FUJITA
156 Twistors in mathematics and physics, T.N. BAILEY & R.J. BASTON (eds)
157 Analytic pro-*p* groups, J.D. DIXON, M.P.F. DU SAUTOY, A. MANN & D. SEGAL
158 Geometry of Banach spaces, P.F.X. MÜLLER & W. SCHACHERMAYER (eds)
159 Groups St Andrews 1989 volume 1, C.M. CAMPBELL & E.F. ROBERTSON (eds)
160 Groups St Andrews 1989 volume 2, C.M. CAMPBELL & E.F. ROBERTSON (eds)
161 Lectures on block theory, BURKHARD KÜLSHAMMER
162 Harmonic analysis and representation theory, A. FIGA-TALAMANCA & C. NEBBIA
163 Topics in varieties of group representations, S.M. VOVSI
164 Quasi-symmetric designs, M.S. SHRIKANDE & S.S. SANE
165 Groups, combinatorics & geometry, M.W. LIEBECK & J. SAXL (eds)
166 Surveys in combinatorics, 1991, A.D. KEEDWELL (ed)
167 Stochastic analysis, M.T. BARLOW & N.H. BINGHAM (eds)
168 Representations of algebras, H. TACHIKAWA & S. BRENNER (eds)
169 Boolean function complexity, M.S. PATERSON (ed)
170 Manifolds with singularities and the Adams-Novikov spectral sequence, B. BOTVINNIK
171 Squares, A.R. RAJWADE
172 Algebraic varieties, GEORGE R. KEMPF
173 Discrete groups and geometry, W.J. HARVEY & C. MACLACHLAN (eds)
174 Lectures on mechanics, J.E. MARSDEN
175 Adams memorial symposium on algebraic topology 1, N. RAY & G. WALKER (eds)
176 Adams memorial symposium on algebraic topology 2, N. RAY & G. WALKER (eds)
177 Applications of categories in computer science, M. FOURMAN, P. JOHNSTONE, & A. PITTS (eds)
178 Lower K- and L-theory, A. RANICKI
179 Complex projective geometry, G. ELLINGSRUD *et al*
180 Lectures on ergodic theory and Pesin theory on compact manifolds, M. POLLICOTT
181 Geometric group theory I, G.A. NIBLO & M.A. ROLLER (eds)
182 Geometric group theory II, G.A. NIBLO & M.A. ROLLER (eds)
183 Shintani zeta functions, A. YUKIE
184 Arithmetical functions, W. SCHWARZ & J. SPILKER
185 Representations of solvable groups, O. MANZ & T.R. WOLF
186 Complexity: knots, colourings and counting, D.J.A. WELSH
187 Surveys in combinatorics, 1993, K. WALKER (ed)
188 Local analysis for the odd order theorem, H. BENDER & G. GLAUBERMAN
189 Locally presentable and accessible categories, J. ADAMEK & J. ROSICKY
190 Polynomial invariants of finite groups, D.J. BENSON
191 Finite geometry and combinatorics, F. DE CLERCK *et al*
192 Symplectic geometry, D. SALAMON (ed)
193 Computer algebra and differential equations, E. TOURNIER (ed)
194 Independent random variables and rearrangement invariant spaces, M. BRAVERMAN
195 Arithmetic of blowup algebras, WOLMER VASCONCELOS
196 Microlocal analysis for differential operators, A. GRIGIS & J. SJÖSTRAND
197 Two-dimensional homotopy and combinatorial group theory, C. HOG-ANGELONI, W. METZLER & A.J. SIERADSKI (eds)
198 The algebraic characterization of geometric 4-manifolds, J.A. HILLMAN
199 Invariant potential theory in the unit ball of $\mathbf{C}^n$, MANFRED STOLL
200 The Grothendieck theory of dessins d'enfant, L. SCHNEPS (ed)
201 Singularities, JEAN-PAUL BRASSELET (ed)
202 The technique of pseudodifferential operators, H.O. CORDES
203 Hochschild cohomology of von Neumann algebras, A. SINCLAIR & R. SMITH
204 Combinatorial and geometric group theory, A.J. DUNCAN, N.D. GILBERT & J. HOWIE (eds)
205 Ergodic theory and its connections with harmonic analysis, K. PETERSEN & I. SALAMA (eds)
206 Lectures on noncommutative geometry, J. MADORE
207 Groups of Lie type and their geometries, W.M. KANTOR & L. DI MARTINO (eds)
208 Vector bundles in algebraic geometry, N.J. HITCHIN, P. NEWSTEAD & W.M. OXBURY (eds)
209 Arithmetic of diagonal hypersurfaces over finite fields, F.Q. GOUVÊA & N. YUI
210 Hilbert C*-modules, E.C. LANCE
211 Groups 93 Galway / St Andrews I, C.R. CAMPBELL *et al*
212 Groups 93 Galway / St Andrews II, C.R. CAMPBELL *et al*
214 Generalised Euler-Jacobi inversion formula and asymptotics beyond all orders, V. KOWALENKO, N.E. FRANKEL, M.L. GLASSER & T. TAUCHER
215 Number theory, S. DAVID (ed)
216 Stochastic partial differential equations, A. ETHERIDGE (ed)
217 Quadratic forms with applications to algebraic geometry and topolgy, A. PFISTER
218 Surveys in combinatorics, 1995, PETER ROWLINSON (ed)
221 Harmonic approximation, S.J. GARDINER
222 Advances in linear logic, J.-Y. GIRARD, Y. LAFONT & L. REGNIER (eds)

London Mathematical Society Lecture Note Series. 214

Generalised Euler-Jacobi inversion formula and asymptotics beyond all orders

V. Kowalenko
University of Melbourne

N. E. Frankel
University of Melbourne

M. L. Glasser
Clarkson University, New York

T. Taucher
Formerly University of Melbourne

CAMBRIDGE UNIVERSITY PRESS
Cambridge, New York, Melbourne, Madrid, Cape Town,
Singapore, São Paulo, Delhi, Mexico City

Cambridge University Press
The Edinburgh Building, Cambridge CB2 8RU, UK

Published in the United States of America by Cambridge University Press, New York

www.cambridge.org
Information on this title: www.cambridge.org/9780521497985

First published 1995

A catalogue record for this publication is available from the British Library

Library of Congress Cataloguing in Publication Data

ISBN 978-0-521-49798-5 Paperback

PREFACE

One of the major aims of this work is to derive inversion formulae for a number theoretic series, which we call the generalised Euler-Jacobi series, and other related series. We write this series as

$$S_{p/q}(a) = \sum_{n=0}^{\infty} e^{-an^{p/q}} \quad ,$$

where p and q are integers, and a is a positive real number. The series studied in this work appear in Chapter 15 of *Ramanujan's Notebooks Part II* by B.C. Berndt [1] together with asymptotic expansions as $a \to 0^+$, but as we shall see, these asymptotic expansions are, in many instances, vacuous. By using our inversion formulae, however, we aim to develop more complete asymptotic expansions than those obtained by Ramanujan and Berndt. The beauty of the inversion formulae is that they transform these series slowly converging in the small parameter a into ones that are rapidly converging. Jacobi's theta function series, recovered here for $p/q = 2$, is the classic example. Our next major aim is to exhibit by way of known hypergeometric function theory, transcendental exponential series that cannot be evaluated by employing the canonical Mellin transform technique. For the case where p/q is an even integer we shall see that these exponential terms become the dominant asymptotic series, but for all other values of $p/q > 2$ they are subdominant to the algebraic series obtained by Ramanujan and Berndt.

Our motivation for evaluating transcendental exponential expressions of the form $\exp(-1/\epsilon)$ in the asymptotic expansions presented in this work has been spurred by the exciting developments occurring in asymptotic analysis over the last decade. Since each term occurring in the conventional Poincaré definition of an asymptotic series is algebraic, transcendental exponential terms are not captured by the series in the limit as $\epsilon \to 0^+$ and as a consequence, these small terms have been traditionally neglected in asymptotics. They are said to lie beyond all orders of an asymptotic expression.

Over the last decade outstanding problems in dendritic crystal growth, the directional solidification of crystals, viscous flows in the presence/absence of surface tension, quantum tunnelling, ordinary differential equations and optics have required improved methods designed to obtain meaningful corrections that lie beyond all orders of a conventional asymptotic expansion. For these exceptional problems, conventional asymptotic analysis is simply inadequate and so the new subject of asymptotics beyond all orders has evolved. This subject refers to the collection of methods used to derive transcendentally small terms hiding behind all orders of a divergent asymptotic series. For a more detailed account of these methods and their role in the problems presented above, the reader is directed to the recent book by Segur et al [2].

The need to retain exponential corrections becomes apparent when one considers small but finite values of ϵ. This is essentially the domain where the practical use of an asymptotic expansion occurs. For example, $\exp(-1/\epsilon)$ is smaller than any power of ϵ as $\epsilon \to 0^+$ but for $\epsilon = 1/2$, $\exp(-1/\epsilon) > \epsilon^3$ and $10e^{-1/\epsilon} > 1$. As a consequence, transcendentally small terms can be numerically important in practical applications, especially when it is realised that as ϵ moves away from 0^+, more and more terms are required to match the conventional asymptotic series with the exact solution. These may not be necessarily algebraic because as ϵ moves away from 0^+ the optimal number of algebraic terms before the asymptotic series begins to diverge may decrease. Thus, there is no other method of approaching the original function but to include these transcendental exponential terms. We shall see this more clearly when we discuss asymptotics beyond all orders in relation to the generalised Euler-Jacobi series in Sec. 7.

As explained in the prologue of his book [3], Dingle, in a series of papers in the late 1950s, was able to derive and evaluate integral representations for the remainder terms appropriate to a large number of important alternating and single-sign asymptotic series just from the general term in each series. These demonstrations meant that asymptotic series could be precisely interpreted despite their ultimate divergence and, as a consequence, implied that Poincaré's definition had to be either supplemented or replaced. As discussed above, literal adherence to this definition means that exponentially decreasing functions of the variable are to be discarded. Yet, if such terms are to be discarded, then the abridged asymptotic expansion will not, of course, reproduce the original function. According to Dingle, this theoretic failure in uniqueness is solely doctrinaire, the result of an over-permissive prescription; in actual practice, a sufficiently detailed analysis for a particular function in a phase sector will yield its 'complete asymptotic expansion' [3,4] including any sets of exponentially small terms which may be present. Thus, by 1959 there was more than a glimpse of an exciting prospect. If ways of deriving asymptotic expansions could be extended so as to yield not only the usual exact terms for the first few terms but also an accurate expression for the general late term enabling the divergent part to be interpreted, then asymptotics would liberate itself from its earlier drawbacks of vagueness and concomitant severe limitation in accuracy and range in applicability, and be elevated to a discipline yielding precise answers.

As theoretical physicists we, like Dingle, are intrigued by this prospect since it will enable us to handle more easily the large variety of mathematical functions which abound in our field. It would also establish a definite link between the asymptotic approach in mathematics and the theoretician's criteria for suitable methods to tackle physical problems. It is by no means a mere accident that the world of physics turns out to be, on close examination, asymptotic rather than convergent. As an example, only a more highly developed theory of asymptotics could decisively clear up the two crucial challenges to the current perturbation techniques which in effect rearrange terms so that the first few terms form a decreasing sequence without actually guaranteeing a similar improvement in later orders. A more highly developed theory of asymptotics would allow us to get closer to the exact function, which is our ultimate aim in this work. Such a theory would have a profound effect in physics, thereby removing the need to renormalise expansions by subtracting infinities to obtain a finite result, which has always been a bane to mathematicians.

Dingle states that the four main tasks in creating a new theory of asymptotics are:

(i) to elucidate the origin and nature of asymptotic expansions and by utilising this understanding to construct definitions which avoid ambiguities and the concomitant inexactitude of Poincaré's definition.

(ii) to formulate methods of deriving asymptotic expansions from convergent series, integral representations and second order linear differential equations in such a way that exponentially small parts do not get lost and that a substantial number of terms in each component series can be found with ease.

(iii) to extend these methods to provide an expression for the general late term in an asymptotic expansion.

(iv) to develop a systematic theory for interpreting asymptotic expansions beyond their least term.

As we shall see, our study of the generalised Euler-Jacobi series will touch upon all four tasks.

In his pursuit of Dingle's goal of a new asymptotic theory, Berry [5] has fashioned the terms superasymptotics and hyperasymptotics. Superasymptotics constitutes asymptotics beyond all orders, since truncation of an asymptotic series occurs not at a fixed order, following Poincaré, but at the least term thereby achieving small exponential errors of the form of $\exp(-1/\epsilon)$ as $\epsilon \to 0^+$. Hyperasymptotics refers to further improvements to the exponentially small remainder of an optimally truncated series and thus goes 'beyond asymptotics beyond all orders'. Berry's study of divergent series using superasymptotics and hyperasymptotics is founded upon two ideas. He states

> First, that an asymptotic series is a compact encoding of a function and its divergence should be regarded not as a deficiency but as a source of information about the function. In particular, divergence usually indicates the presence of exponentially small terms (but not always as we shall see later in this work) which the bare asymptotic series, uninterpreted, cannot capture. This is why superasymptotics can yield exponential accuracy. A consequence is that the late terms of the asymptotic series associated with one exponential are frequently related by 'resurgence' to the early terms of the asymptotic series associated with another exponential. Second, that the divergences of the series... follow

a common pattern; factorial divided by a power. Recognition of this universality and its cause leads to powerful resummation techniques enabling the asymptotics to be decoded to yield precise (hyperasymptotic) numerical information. These principles were systematically explored and exploited by Dingle in the 1950s, and summarised in his 1973 book, but are only now becoming widely known.

Berry has been able to extend Dingle's work in two ways. For each method the knowledge/evaluation of the transcendental exponential terms appearing in an asymptotic expansion is required. The first method pertains to the Stokes phenomenon, which is characterised by rapid jumps occurring in the multiplier of a small (subdominant) exponential as a result of changing the phase of one or more of the parameters of an asymptotic expansion. According to Berry, such changes in form necessarily accompany the divergence of asymptotic series associated with each exponential, reflecting its inability to describe the other exponentials. By appropriate magnification and resummation, however, a precise description of the change in the multiplier of the subdominant terms can be obtained, which represents a refinement of the phenomenon and the first stage of hyperasymptotics. Thus the evaluation of subdominant exponential terms is crucial when developing asymptotic expansions in the complex plane, mainly because across anti-Stokes lines the subdominant terms become dominant while the dominant terms recede. Although in this work we shall only be concerned with real values of the parameter a in the generalised Euler-Jacobi series, we shall conclude by showing how the results presented here can be extended into the complex plane and shall give exact results for the specific cases of $p/q = 2$ and $p/q = 3$ in terms of special functions. Obtaining asymptotic expansions for the latter case will then require taking into account the Stokes phenomenon which we reserve for future study.

The second advance made by Berry goes much further. By exploiting resurgence, the remainder in an optimally truncated expansion can itself be expressed as an asymptotic series, which in turn has its own remainder. This second remainder can also be optimally truncated with its remainder being expressed as an asymptotic series. The process is then continued producing an intricate sequence of hyperseries in which the original asymptotic coefficients are renormalised by certain universal functions given by n-fold multiple integrals [6], which arise after Borel summation. For these integrals to converge, successive hyperseries must contain fewer terms, thereby guaranteeing that the process of hyperasymptotics must eventually come to a halt. For integrals of exponentials involving several saddle points, Berry and Howls [7] have obtained a notable resurgence identity connecting the expansions about different saddle points. According to Berry [5], this can be employed to refine the method of steepest descent into an exact technique, whose hyperasymptotics can be carried out without the need to resum divergent series.

It is the main purpose of this work to link a fundamental problem in classical analysis with the newly emerging subject of asymptotics beyond all orders, thereby reviving the old and providing the new with a firmer base. As a consequence of our study, we shall present many new and interesting results. However, important questions will still remain unanswered. The first is whether there are further subdominant exponential terms, which are not available by utilising the present asymptotic theory of hypergeometric functions given in Luke [8]. For his part, Luke states that the asymptotic theory he gives is complete and the results presented in Sec. 7 of this work are consistent with this statement but the question is perhaps best resolved by returning to the seminal work of Braaksma [9] to see whether or not the Berry and Howls hyperasymptotic approach can yield these hypothetical 'sub-subdominant' terms. In their paper dealing with the hyperasymptotics of integrals with multiple saddle points, i.e. Ref. [7], Berry and Howls state that the divergences in the asymptotic expansions for these integrals are due to the existence of other saddle points, through which the path of steepest descent does not pass. It is these saddle points that produce the exponentially small terms in an asymptotic expansion. Thus, they refine the method of steepest descent to include the contributions from these points, thereby extending the accuracy of the method. In his study of the asymptotics of hypergeometric functions, Braaksma has stated that the estimation of the remainder terms is the most difficult part, but has sketched how the method of steepest descent can be employed to estimate these terms. It may therefore be possible to improve his results by incorporating the refinement of the method put forward by Berry and Howls, thereby answering the question of whether there are further sub-subdominant terms. If such terms do not arise, then we can be assured that Luke's asymptotic expansions are indeed complete.

The second question is whether we can devise an appropriate methodology for handling the recursion relations that arise when utilising the asymptotic theory of hypergeometric functions to evaluate the generalised Euler-Jacobi series. As we shall see in Sec. 7, in order to apply Dingle's theory of terminants to the asymptotic results presented in this work, we shall need to know how the coefficients in each asymptotic series comprising the complete asymptotic expansion behave. This in turn requires that the solutions to the recursion relations be known or evaluated. Although we shall present recursion relations for various p/q values of the generalised Euler-Jacobi series, we shall only be able to determine the exact behaviour of the coefficients in each asymptotic series of the complete asymptotic expansion when $p/q = 3$. Hence, our analysis in Sec. 7 will be confined to applying Dingle's theory of terminants to $S_3(a)$ and to a short discussion concerning the evaluation of the recursion relations for other values of p/q.

Another issue is whether an appropriate theory of divergent series based on the material in Secs. 5,7 and 8 can be devised since a complete understanding of all asymptotic expansions is intimately connected with understanding divergent series. Only with the development of such a theory can asymptotics be elevated from a subject with deficiencies in accuracy and range of application to one yielding precise results over domains previously thought to be invalid. The issues raised above represent an ambitious programme that goes well beyond the level of the present work and are to be investigated in the future.

ACKNOWLEDGEMENTS

The authors are grateful to the Australian Research Council (ARC) for its support of this work. One of us (M.L.G.) thanks Melbourne University for its hospitality. We thank Andrew Logothetis for his willingness and perseverance in preparing this difficult manuscript. We are grateful to Professor J. Boersma, Eindhoven University of Technology, The Netherlands, for a critical reading of this work and for pointing out some corrections. We also thank Carl Dettmann for his advice on some numerical aspects of this work.

Contents

1. INTRODUCTION

The imaginary transformation, $q(t)$ to $q(it)$, enables the Jacobi theta function $\theta_3(z|q(t))$ to be recast as

$$\theta_3(z|q(it)) = \sum_{n=-\infty}^{\infty} e^{-n^2t+2inz} \quad , \tag{1.1}$$

where z is any arbitrary complex variable and the complex variable t obeys the condition $\mathrm{Re}\, t > 0$. In this form Bellman [10] states that the theta function plays a crucial role in the development of the theory of linear partial differential equations of parabolic type, in the study of the Riemann zeta function, $\zeta(s)$, in connection with the representation of numbers as sums of squares and especially in the theory of elliptic functions.

Jacobi studied Eq. (1.1) in a thorough fashion; the $z = 0$ case yields the remarkable and now most familiar transformation formula [10,11]

$$S_2(a) = \sum_{n=0}^{\infty} e^{-an^2} = \frac{1}{2}\sqrt{\frac{\pi}{a}} + \frac{1}{2} + \sqrt{\frac{\pi}{a}} \sum_{n=1}^{\infty} e^{-n^2\pi^2/a} \quad , \ \mathrm{Re}\ a > 0. \tag{1.2}$$

The advantage of the above transformation is that both the small and large a values of the series, i.e. the asymptotic behaviour, can be evaluated after considering only a limited number of terms. In this work our aim is to investigate the nature of such transformations for the generalised Euler-Jacobi series, which we write as

$$S_{p/q}(a) = \sum_{n=0}^{\infty} e^{-an^{p/q}} \quad , \tag{1.3}$$

where p and q are relatively prime integers. After evaluating a general result, we shall examine specific cases, which are also of interest in statistical mechanics. Although p and q may have common factors, we shall see from what follows that our results will be considerably simplified if p/q is proper.

Berndt [1] notes that Ramanujan derived asymptotic expressions as $a \to 0^+$ for the following series:

$$S^r_{p/q}(a) = \sum_{n=1}^{\infty} e^{-an^{p/q}} n^{r-1} \quad , \ \mathrm{Re}\ r \geq 1 \ , \tag{1.4}$$

and

$$S'(a) = \sum_{n=1}^{\infty} e^{-an} \log n \quad . \tag{1.5}$$

Eq. (1.4) is a further generalisation of our Euler-Jacobi series while Eq. (1.5) is just a variant of Eq. (1.4) since $S' = (d/dr) S^r_1 |_{r=1}$. In Theorem 3.1 of Ch.

15 in Ref. [1], Berndt shows that Eq. (1.4) can be represented as a sum of Riemann zeta functions. However, he failed to point out that for negative even integers the zeta function vanishes, thereby indicating that his result can become vacuous. For example, for $p/q = 2$ and $r = 1$, Eq. (1.4) reduces to Eq. (1.2); yet the result given by Berndt would not produce the series on the right hand side (r.h.s.) of Eq. (1.2), which also contradicts the result given in a corollary to Entry 7 in Ch. 14 of Ref. [1]. In this work we supply the exponential corrections, which are missing from Ramanujan's series for $p/q \geq 2$.

This work is organised as follows. In Sec. 2 we derive the exact transformation for the series $S^r_{p/q}(a)$. Although the series

$$S^{r,m}_{p/q}(a) = \sum_{n=1}^{\infty} e^{-an^{p/q}} n^{r-1} (\log n)^m \quad , \text{ Re } r \geq 1 \ , \tag{1.6}$$

with m a natural number can be expressed as an m-fold derivative of $S^r_{p/q}(a)$ with respect to (w.r.t.) r, it can only be evaluated for $m = 1$. We then put $r = 1$ and $m = 0$ in order to evaluate the transformation for the generalised Euler-Jacobi series Eq. (1.3). Utilising this result, we develop further extensions of it for special cases. In Sec. 3 we begin our study of the asymptotic behaviour for the generalised Euler-Jacobi series by identifying properties of the results presented in Sec. 2. Then we explain how the missing exponential terms in the Ramanujan-Berndt result arise and describe the cancellation of all the growing terms. We also show that the Ramanujan-Berndt result becomes the dominant contribution as $a \to 0^+$ for p/q equal to an odd integer, but does not contribute for even integers. In Sec. 4, we examine the asymptotic expansion for the generalised Euler-Jacobi series when p/q equals 3 by using the method of steepest descent, which will serve as a useful check when we evaluate the asymptotic series by using the inversion formula presented in Sec. 2. In addition, we discuss the problems in applying the method of steepest descent to larger integer values of p/q. In Sec. 5, we evaluate the asymptotic series for specific values of p/q greater than 0 and less than 2. In evaluating the series for p/q equal to 4/3 and 3/2, we encounter the Stokes phenomenon, which is used to show that there are no exponential corrections at these values. The next section contains the detailed evaluation of the a-asymptotic series for all integers up to and including p/q equal to 7. This extensive set of p/q values has been chosen and studied in detail in order to display the rich variety of asymptotic forms exhibited by the generalised Euler-Jacobi series. Whilst we have been able to refine many of the calculational steps for successively higher integers, we shall describe the additional complications that arise for higher integers. These make the calculation of the asymptotic series $a \to 0^+$ beyond p/q equal to 7 a somewhat tedious and laborious task. Nevertheless, an appropriate methodology is presented, should the interested reader desire to calculate the asymptotic behaviour of the generalised Euler-Jacobi series for p/q values other than those presented in this work or for various values of r and m in the series given by Eq. (1.6). In Sec. 7 we apply Dingle's theory of terminants to the asymptotic expansion for $S_3(a)$ obtained in Sec. 6 and show by including the subdominant exponential series that the range of applicability of an asymptotic expansion can be extended away from the expansion parameter's limit point, in our case $a \to 0^+$, to the intermediate region, i.e. $a \geq O(1)$. Thus we give meaning to the inclusion of all subdominant asymptotic series in an asymptotic expansion. We then describe a methodology for evaluating error estimates for all the subdominant exponential series presented in Sec. 6 by concentrating on the exponential series for $S_4(a)$. We also show how the remainders for the algebraic series that arise when p/q is not equal to an even integer can be expressed exactly in terms of convergent integral representations by using Dingle's theory of terminants. Although the various convergent integrals can be evaluated numerically using standard routines, in order to obtain the extremely high level of accuracy in an expedient manner required for the study in Sec. 7, we had to devise an alternative approach for their evaluation. We call this the Mellin-Barnes regularisation procedure and describe it in detail in Sec. 8. In Sec. 9 we discuss the results obtained in the earlier sections and outline further extensions of the material presented in this work.

2. EXACT EVALUATION OF $S^r_{p/q}(a)$

In this section we present the exact evaluation of Eq. (1.6) and then give the exact answer for the simpler generalised Euler-Jacobi series. Our approach is different from that employed by Berndt in Ref. [1] and as a consequence, our answer will be expressed in terms of sums of hypergeometric series as opposed to the results given by the others in terms of sums of the Riemann zeta function. However, before introducing our approach, we shall briefly review the one Berndt used to obtain Ramanujan's answer for $S_{p/q}(a)$ and then shall discuss the limitations of the result.

A simple integral representation for $S^r_{p/q}(a)$ can be obtained by evaluating and inverting the Mellin transform with respect to a. This gives

$$S^r_{p/q}(a) = \frac{1}{2\pi i} \int_{\sigma-i\infty}^{\sigma+i\infty} ds\, \Gamma(s)\, \zeta\left(1 - r + \frac{ps}{q}\right) a^{-s} \ , \tag{2.1}$$

where $\sigma = \text{Re}\ s > \sup\{0, qr/p\}$. Berndt then considers a contour integral involving the above integrand around a positively oriented rectangle with vertices $\sigma \pm iT$ and $-M \pm iT$, where $T > 0$ and $M = N + \frac{1}{2}$. The integer N is chosen to be sufficiently large that $N > |r|q/p$. Then the integrand has simple poles at $s = qr/p$ and $s = 0, -1, \cdots, -N$, unless $qr/p = -k \in \mathbf{Z}^-$, in which case a double pole exists at $s = -k$. By applying the residue theorem to the contour integral denoted by $I_{M,T}$, Berndt obtains

$$I_{M,T} = \frac{q\,\Gamma(qr/p)}{p\,a^{qr/p}} + \sum_{l=0}^{N} \frac{(-a)^l}{l!}\,\zeta\left(1 - r - \frac{pl}{q}\right) , \tag{2.2}$$

for $qr/p \notin \mathbf{Z}^-$, whereas if $qr/p = -k \in \mathbf{Z}^-$, then

$$I_{M,T} = \left\{\frac{q}{p}(H_r - \gamma) + \gamma - \frac{q}{p}\log a\right\}\frac{(-a)^k}{k!} + \sum_{l=0}^{N}{}^{*} \frac{(-a)^l}{l!}\,\zeta\left(1 - r - \frac{pl}{q}\right) . \tag{2.3}$$

In Eq. (2.3), the asterisk in the summation means that those values of l yielding undefined summands are excluded, while γ denotes Euler's constant. In addition, $H_r = \sum_{l=1}^{r} l^{-1}$.

If the integral is evaluated around the rectangle, then it is found that the two contributions parallel to the real axis vanish as $T \to \infty$, leaving the two contributions parallel to the imaginary axis. One of these is just the integral in Eq. (2.1), whereas the remainder is shown to be much less than a^M as $a \to 0^+$. Hence, Berndt shows that $S^r_{p/q}(a)$ approaches the results given by Eqs. (2.2) and (2.3), depending on the values of qr/p.

If we put $r = 1$ in Eq. (2.1), so that we get the generalised Euler-Jacobi series, then according to Berndt

$$S_{p/q}(a) \sim \frac{q\,\Gamma(q/p)}{p\,a^{q/p}} + \frac{1}{2} + \sum_{l=1}^{\infty} \frac{(-a)^l}{l!}\,\zeta\left(-\frac{pl}{q}\right) . \tag{2.4}$$

Now if p/q equals an even integer, then the sum in Eq. (2.4) vanishes because the trivial zeros of the Riemann zeta function are the negative even integers. As mentioned earlier, putting p/q equal to 2 in the above equation does not give the exponential series on the r.h.s. of Eq. (1.2). Therefore, since the remainder integral is much less than a^M as $a \to 0^+$, one can only conclude that the missing terms in Eqs. (2.2) and (2.3) are exponential in nature and although small, they become important when p/q is an even integer. On the other hand, if p/q is set equal to unity, then we find

$$S_1(a) \sim 1 + \frac{1}{a} + \sum_{n=0}^{\infty}(-1)^n\,\frac{a^{2n+1}}{(2n+2)!}\,B_{n+1} = (1 - e^{-a})^{-1} , \tag{2.5}$$

where we have expressed the zeta function values in terms of Bernoulli numbers, and used the result on p. 127 of Whittaker and Watson [11] to obtain the final exponential expression. Although the above result is written as asymptotic, it is the exact answer that is obtained by putting p/q equal to unity in Eq. (1.2) and then summing the resultant geometric series. So, whilst it is our ultimate aim to evaluate the small a-asymptotic form of the generalised Euler-Jacobi series for various values of p/q, we shall also find that the missing exponential terms occur for values of $p/q \geq 2$.

To obtain an exact answer for $S^{r,m}_{p/q}(a)$, we note that this series can be expressed as an m-fold derivative of Eq. (1.4) w.r.t. r. Hence, we need only concentrate for the moment on evaluating the exact expression for $S^r_{p/q}(a)$, which, after application of the Euler-Maclaurin summation formula [12], becomes

$$S^r_{p/q}(a) = \frac{q\,\Gamma(qr/p)}{p\;a^{qr/p}} + \frac{1}{2}\,\delta_{r,1} + \frac{1}{\pi}\sum_{n=1}^{\infty}\frac{1}{n}\int_0^{\infty} dt\; e^{-at}\sin\Bigl(2n\pi t^{q/p}\Bigr) \times \left(at^{q(r-1)/p} - \frac{q(r-1)}{p}\,t^{q(r-1)/p-1}\right) \quad . \tag{2.6}$$

The integrals in Eq. (2.6) are all of the form

$$I = \int_0^{\infty} dt\; e^{-at} t^{\alpha} \sin\Bigl(\beta t^{q/p}\Bigr) \quad , \tag{2.7}$$

where Re $a >$ Im β and $\mathrm{Re}\,(\alpha + q/p) > -1$. They can be evaluated by expanding $\sin\left(\beta t^{q/p}\right)$ into its Maclaurin series and then integrating term-by-term. We find

$$I = \sum_{k=0}^{\infty} \frac{(-1)^k \beta^{2k+1}}{a^{(2k+1)q/p+\alpha+1}} \frac{\Gamma\Bigl((2k+1)q/p + \alpha + 1\Bigr)}{\Gamma(2k+2)} \quad . \tag{2.8}$$

We note that the above series is absolutely convergent for $p/q > 1$ and for $p/q = 1$ provided that $\beta/a < 1$. We shall see in our analysis of the case $p/q = 1$ that our results can be continued analytically to values of $\beta/a > 1$. For values of $p/q < 1$ the series is divergent, but is not as divergent as a Gevrey series, though the above series can be expressed as a definite integral of a Gevrey series. In the section dealing with $p/q < 1$ we shall utilise the technique of Borel summation as described by Hardy [13] to recast the original series for each p/q value studied, into convergent integrals which can then be evaluated by various techniques. It should also be noted that Ramis [14] has developed a resummation theory for Gevrey series in his study of coefficient growth in formal power series satisfying a given analytic linear differential equation, whilst Thomann has shown in Ref. [15] how to apply this theory to solutions of complex ordinary differential equations in the neighbourhood of irregular singularities, by various methods incorporating computer algebra. Since resummation techniques can be applied to Gevrey series, it should, therefore, not be surprising that the alternating divergent series given above can also be resummed to yield convergent results for $p/q < 1$.

If we now make the substitution $k = mp + l$ with $p > 1$, so that the above summation over k degenerates into two separate sums, then Eq. (2.8) becomes

$$I = \frac{1}{2}\sum_{l=0}^{p-1}\sum_{j=0}^{\infty}(1+(-1)^j)\, e^{i\pi(jp/2+l)}\, \frac{e^{2ijs\pi}\,\beta^{jp+2l+1}}{a^{jq+(2l+1)q/p+\alpha+1}} \times \frac{\Gamma(jq+(2l+1)q/p+\alpha+1)}{\Gamma(jp+2l+2)}\ , \tag{2.9}$$

where the substitution $j = 2m$ has also been made and s is an arbitrary integer. By applying Gauss' multiplication formula (No. 8.335 of Ref. [16]) we recast the ratio of the two gamma functions in Eq. (2.9) as

$$\frac{\Gamma(jq+(2l+1)q/p+\alpha+1)}{\Gamma(jp+2l+2)} = (2\pi)^{(p-q)/2}\, \frac{q^{jq+(2l+1)q/p+\alpha+1/2}}{p^{jp+2l+3/2}} \times \prod_{n=0,m=0}^{q-1,p-1} \frac{\Gamma(j+(2l+1)/p+(\alpha+1)/q+n/q)}{\Gamma(j+(2l+2)/p+m/p)}\ . \tag{2.10}$$

If Eq. (2.10) is introduced into Eq. (2.9), then the two summations over j can be readily identified as generalised hypergeometric series. Hence, Eq. (2.9) can finally be written as

$$I = \frac{\beta}{2}\, \frac{(2\pi)^{(p-q)/2}\, q^{q/p+\alpha+1/2}}{p^{3/2}\, a^{q/p+\alpha+1}} \sum_{l=0}^{p-1}(-1)^l \left(\frac{\beta\, q^{q/p}}{p\, a^{q/p}}\right)^{2l} \prod_{\Gamma}^{q,p}\left(\frac{2l+1}{p},\frac{\alpha}{q}\right) \times \Bigg[{}_{q+1}F_p\Big(1, \Delta_{q-1}\big(q,\tau(1,\alpha)\big); \Delta_{p-1}\big(p,2l+2\big); \frac{q^q\beta^p}{a^q p^p} e^{i\pi(p/2+2s)}\Big) + {}_{q+1}F_p\Big(1, \Delta_{q-1}\big(q,\tau(1,\alpha)\big); \Delta_{p-1}\big(p,2l+2\big); \frac{q^q\beta^p}{a^q p^p} e^{i\pi(p/2+2s+1)}\Big)\Bigg]\ , \tag{2.11}$$

where

$$\prod_{\Gamma}^{q,p}(\alpha,\beta) = \prod_{n=1,m=1}^{q,p} \frac{\Gamma(\alpha+\beta+n/q)}{\Gamma(\alpha+m/p)}\ , \tag{2.12}$$

$\tau(x,y) = (2l+x)q/p+y+1$ and $\Delta_l(k,x) \equiv \{x/k, (x+1)/k, \cdots, (x+l)/k\}$. To evaluate $S^r_{p/q}(a)$, β must be put equal to $2n\pi$ and α must take the values of $q(r-1)/p$ and $q(r-1)/p-1$, respectively for the first and second integrals in Eq. (2.6). Thus, $S^r_{p/q}(a)$ becomes

$$S^r_{p/q}(a) = \frac{q}{p}\, \frac{\Gamma(qr/p)}{a^{qr/p}} + \frac{1}{2}\,\delta_{r,1} + (2\pi)^{(p-q)/2}\, \frac{q^{q/p+1/2}}{p^{3/2}\, a^{q/p}} \times \sum_{n=1}^{\infty}\sum_{l=0}^{p-1}(-1)^l \left(\frac{2n\pi\, q^{q/p}}{p\, a^{q/p}}\right)^{2l} \left(\frac{q}{a}\right)^{q(r-1)/p} \Bigg[\prod_{\Gamma}^{q,p}\left(\frac{2l+1}{p},\frac{r-1}{p}\right)$$

$$\times \; {}_{q+1}F_p\Big(1, \Delta_{q-1}\big(q, \tau(r,0)\big); \Delta_{p-1}\big(p, 2l+2\big); \pm z e^{i\pi(p/2+2s)}\Big)^+$$
$$- \left(\frac{r-1}{p}\right) \prod_{\Gamma}^{q,p}\left(\frac{2l+1}{p}, \frac{r-1}{p} - \frac{1}{q}\right)$$
$$\times \; {}_{q+1}F_p\Big(1, \Delta_{q-1}\big(q, \tau(r,0)\big); \Delta_{p-1}\big(p, 2l+2\big); \pm z e^{i\pi(p/2+2s)}\Big)^+\Bigg] , \qquad (2.13)$$

where $\delta_{r,1}$ is the Kronecker delta and z equals $(q/a)^q(2n\pi/p)^p$. In the above we have also introduced the following notation, which will be used throughout this work:

$$ {}_qF_p(\alpha_1, \cdots, \alpha_q; \beta_1, \cdots, \beta_p; \pm z)^{\pm} = {}_qF_p(\alpha_1, \cdots, \alpha_q; \beta_1, \cdots, \beta_p; z)$$
$$\pm \; {}_qF_p(\alpha_1, \cdots, \alpha_q; \beta_1, \cdots, \beta_p; -z) \; .$$

When evaluating asymptotic forms we shall choose s values so that the variables in the hypergeometric functions obey

$$-\pi < \arg\left(\pm z e^{i\pi(p+4s)/2}\right) \leq \pi \; .$$

Putting r equal to unity in Eq. (2.13) gives the exact result for the generalised Euler-Jacobi series, which is

$$S_{p/q}(a) = \frac{q\,\Gamma(q/p)}{p\,a^{q/p}} + \frac{1}{2} + \frac{(2\pi)^{(p-q)/2}\, q^{q/p+1/2}}{p^{3/2}\, a^{q/p}} \sum_{n=1}^{\infty}\sum_{l=0}^{p-1}(-1)^l z^{2l/p} \prod_{\Gamma}^{q,p}\left(\frac{2l+1}{p}, 0\right)$$
$$\times \; {}_qF_{p-1}\Big(1, \Delta_{q-2}\big(q, \tau(1,0)\big); \Delta_{p-2}\big(p, 2l+2\big); \pm\, z e^{i\pi p'/2}\Big)^+ . \qquad (2.14)$$

In Eq. (2.14) p' equals $p+4s$.

We can also use the relationships between hypergeometric functions and Meijer G-functions, which are given in Sec. 9.34 of Gradshteyn and Ryzhik [16], to allow us to express Eqs. (2.13) and (2.14) in simpler notation as

$$S^r_{p/q}(a) = \frac{q\Gamma(qr/p)}{p\,a^{qr/p}} + \frac{1}{2}\,\delta_{r,1} + (2\pi)^{(r-q)/2}\,\frac{q^{q/p+1/2}}{p^{3/2}\,a^{q/p}}$$
$$\times \sum_{n=1}^{\infty}\sum_{l=0}^{p-1}(-1)^l\, z^{2l/p}\left(\frac{q}{a}\right)^{q(r-1)/p}\left(\frac{r-1}{p}\right)\left[\left(\frac{p}{r-1}\right)\right.$$
$$\times\; G^{1,q+1}_{q+1,p+1}\left(\pm\, z e^{i\pi p'/2}\,\middle|\, \begin{matrix} 0, -\frac{2l+r}{p}, \frac{1}{q} - \frac{2l+r}{p}, \cdots, \frac{q-1}{q} - \frac{2l+r}{p} \\ 0, -\frac{2l+1}{p}, -\frac{2l}{p}, \cdots, \frac{p-2l-2}{p} \end{matrix}\right)^+$$
$$\left. - \,G^{1,q+1}_{q+1,p+1}\left(\pm z\, e^{i\pi p'/2}\,\middle|\, \begin{matrix} 0, \frac{1}{q} - \frac{2l+r}{p}, \frac{2}{q} - \frac{2l+r}{p}, \cdots, 1 - \frac{2l+r}{p} \\ 0, -\frac{2l+1}{p}, -\frac{2l}{p}, \cdots, \frac{p-2l-2}{p} \end{matrix}\right)^+\right] , \qquad (2.15)$$

and

$$S_{p/q}(a) = \frac{q\,\Gamma(q/p)}{p\,a^{q/p}} + \frac{1}{2} + (2\pi)^{(p-q)/2}\,\frac{q^{q/p+1/2}}{p^{3/2}\,a^{q/p}}\sum_{n=1}^{\infty}\sum_{l=0}^{p-1}(-1)^l\,z^{2l/p}$$
$$\times\ G^{1,q}_{q,p}\left(\pm z e^{i\pi p'/2}\,\middle|\,\begin{matrix}0, \frac{1}{q}-\frac{2l+1}{p}, \frac{2}{q}-\frac{2l+1}{p}, \ldots, 1-\frac{1}{q}-\frac{2l+1}{p}\\ 0, -\frac{2l}{p}, \frac{1}{p}-\frac{2l}{p}, \ldots, \frac{p-2l-2}{p}\end{matrix}\right)^{+}. \tag{2.16}$$

According to p. 225 of Luke [8], the above results are only valid when $q+1 \le p$ or when $p = q$ and $|z| < 1$.

Alternative representations of Eqs. (2.15) and (2.16) exist for the same conditions on p, q and z. These are, respectively,

$$S^r_{p/q}(a) = \frac{q\,\Gamma(qr/p)}{p\,a^{qr/p}} + \frac{1}{2}\,\delta_{r,1} + (2\pi)^{(p-q)/2}\,\frac{q^{q/p+1/2}}{p^{3/2}\,a^{q/p}}\sum_{n=1}^{\infty}\sum_{l=0}^{p-1}(-1)^l\,z^{2l/p}$$
$$\times\ \left(\frac{q}{a}\right)^{q(r-1)/p}\left[\,G^{q+1,1}_{p+1,q+1}\left(\pm z^{-1}e^{-i\pi p'/2}\,\middle|\,\begin{matrix}1, \frac{2l+2}{p}, \frac{2l+3}{p}, \ldots, \frac{2l+p+1}{p}\\ 1, \frac{2l+r}{p}+\frac{1}{q}, \ldots, \frac{2l+r}{p}+1\end{matrix}\right)^{+}\right.$$
$$\left.-\left(\frac{r-1}{p}\right)G^{q+1,1}_{p+1,q+1}\left(\pm z^{-1}e^{-i\pi p'/2}\,\middle|\,\begin{matrix}1, \frac{2l+2}{p}, \ldots, \frac{2l+p+1}{p}\\ 1, \frac{2l+r}{p}, \frac{2l+r}{p}+\frac{1}{q}, \ldots, \frac{2l+r}{p}+\frac{q-1}{q}\end{matrix}\right)^{+}\right], \tag{2.17}$$

and

$$S_{p/q}(a) = \frac{q\Gamma(q/p)}{p\,a^{q/p}} + \frac{1}{2} + (2\pi)^{(p-q)/2}\,\frac{q^{q/p+1/2}}{p^{3/2}\,a^{q/p}}\sum_{n=1}^{\infty}\sum_{l=0}^{p-1}(-1)^l\,z^{2l/p}$$
$$\times\ G^{q,1}_{p,q}\left(\pm z^{-1}e^{-i\pi p'/2}\,\middle|\,\begin{matrix}1, \frac{2l+2}{p}, \frac{2l+3}{p}, \ldots, \frac{2l+p}{p}\\ 1, \frac{2l+1}{p}+\frac{1}{q}, \frac{2l+1}{p}+\frac{2}{q}, \ldots, \frac{2l+1}{p}+\frac{q-1}{q}\end{matrix}\right)^{+}. \tag{2.18}$$

By taking the derivative of Eq. (2.13) w.r.t. r, we are able to evaluate $S^{r,1}_{p/q}(a)$, which is given by

$$S^{r,1}_{p/q}(a) = \frac{(q/p)^2}{a^{q/p}}\left(\Psi\left(\frac{qr}{p}\right) - \log a\right)\Gamma\left(\frac{qr}{p}\right) + (2\pi)^{(p-q)/2}\,\frac{q^{q/p+3/2}}{p^{5/2}\,a^{q/p}}\log\left(\frac{q}{a}\right)$$
$$\times\ \sum_{n=1}^{\infty}\sum_{l=0}^{p-1}(-1)^l\,z^{2l/p}\left(\frac{q}{a}\right)^{q(r-1)/p}\left[\,{}_{q+1}\mathbf{F}_{p}\left(\frac{2l+1}{p}, \frac{r-1}{p}, \pm z e^{i\pi p'/2}\right)^{+}\right.$$
$$\left.-\left(\frac{r-1}{p} + \frac{1}{q\log(q/a)}\right)\,{}_{q+1}\mathbf{F}_{p}\left(\frac{2l+1}{p}, \frac{r-1}{p}-\frac{1}{q}, \pm z e^{i\pi p'/2}\right)^{+}\right]$$
$$+\ (2\pi)^{(p-q)/2}\,\frac{q^{q/p+1/2}}{p^{3/2}\,a^{q/p}}\sum_{n=1}^{\infty}\sum_{l=0}^{p-1}(-1)^l\,z^{2l/p}\left(\frac{q}{a}\right)^{q(r-1)/p}$$
$$\times\ \left[\frac{d}{dr}\,{}_{q+1}\mathbf{F}_{p}\left(\frac{2l+1}{p}, \frac{r-1}{p}, \pm z e^{i\pi p'/2}\right)^{+} - \left(\frac{r-1}{p}\right)\right.$$
$$\left.\times\ \frac{d}{dr}\,{}_{q+1}\mathbf{F}_{p}\left(\frac{2l+1}{p}, \frac{r-1}{p}-\frac{1}{q}, \pm z e^{i\pi p'/2}\right)^{+}\right], \tag{2.19}$$

where

$$
{}_{\mathbf{q+1}}\mathbf{F}_{\mathbf{p}}(\alpha,\beta,z) = \prod_{\Gamma}^{q,p}(\alpha,\beta) \\
\times\ {}_{q+1}F_p\Big(1,\Delta_{q-1}\big(q,(\alpha+\beta)q+1\big);\Delta_{p-1}\big(p,\alpha+1\big);z\Big) \quad ,
$$

and $\Psi(qr/p)$ is the digamma function. For $q+1 \le p$ or $q=p$ and $|z|<1$, ${}_{\mathbf{q+1}}\mathbf{F}_{\mathbf{p}}(\alpha,\beta,z)$ is essentially a Meijer G-function as can be seen from Eqs. (2.15) - (2.18). Furthermore in Eq. (2.19), we note that

$$
\frac{d}{dr}\,{}_{\mathbf{q+1}}\mathbf{F}_{\mathbf{p}}\left(\frac{2l+1}{p},\frac{r-1}{p},\pm\, ze^{i\pi p'/2}\right) = \frac{q}{p}\sum_{k=0}^{\infty}\left(\pm\, ze^{i\pi p'/2}\right)^k \\
\times\ \prod_{\Gamma}^{q,p}\left(k+\frac{2l+1}{p},\frac{r-1}{p}\right)\left(\Psi\left(qk+\frac{(2l+r)q}{p}+1\right)-\log q\right) \quad , \tag{2.20}
$$

and

$$
\frac{d}{dr}\,{}_{\mathbf{q+1}}\mathbf{F}_{\mathbf{p}}\left(\frac{2l+1}{p},\frac{r-1}{p}-\frac{1}{q},\pm\, ze^{i\pi p'/2}\right) = \frac{q}{p}\sum_{k=0}^{\infty}\left(\pm\, ze^{i\pi p'/2}\right)^k \\
\times\ \prod_{\Gamma}^{q,p}\left(k+\frac{2l+1}{p},\frac{r-1}{p}-\frac{1}{q}\right)\left(\Psi\left(qk+\frac{(2l+r)q}{p}\right)-\log q\right) \quad , \tag{2.21}
$$

where we have used No. 8.365(6) in Ref. [16].

To evaluate $S^{r,m}_{p/q}(a)$ for $m>1$, the $(m-1)$-fold derivative of Eq. (2.19) w.r.t. r must be obtained. This is not such a difficult problem for the first expression on the r.h.s. of Eq. (2.19) since derivatives of the digamma function yield polygamma functions, as described in Ch. 6 of Abramowitz and Stegun [17]. However, successively higher order derivatives become increasingly more difficult to evaluate, especially since the derivatives act on the indices or parameters rather than the variable of the hypergeometric functions appearing in Eq. (2.19). Once Eqs. (2.20) and (2.21) are included, Eq. (2.19) becomes very unwieldy to utilise in the determination of the asymptotics for $S^{r,1}_{p/q}(a)$. Hence, the evaluation of the asymptotics for $S^{r,m}_{p/q}(a)$ is a formidable task if derivatives of Eq. (2.13) w.r.t. r are carried out first. An alternative is to evaluate the asymptotics for $S^r_{p/q}(a)$ and then take the m-fold derivatives. Although we shall primarily be concerned with evaluating the asymptotics for the generalised Euler-Jacobi series in this work, since the asymptotics for $S_1(a)$ and $S_2(a)$ are so well-known (see Eqs. (1.2) and (2.5)), we shall begin our study of these cases in Sec. 5 by considering $S^r_1(a)$ and $S^r_2(a)$. As an aside, we shall also evaluate the asymptotics for $S^{r,1}_1(a)$ and $S^{r,1}_2(a)$ by taking the derivatives of the asymptotic forms for $S^r_1(a)$ and $S^r_2(a)$ w.r.t. r, respectively.

It is interesting to compare Eq. (2.19) with the corresponding result obtained by Berndt [1], which is

$$S^{r,1}_{p/q} \sim \left(\frac{\Gamma'(qr/p) - \Gamma(qr/p)\log a}{(p/q)^2\, a^{qr/p}}\right) - \sum_{n=0}^{\infty} (-1)^n \frac{a^n}{n!}\, \zeta'\left(1 - r - \frac{pn}{q}\right) \quad . \tag{2.22}$$

As expected, the first expression on the r.h.s. of Eq. (2.22) is identical to the corresponding expression in Eq. (2.19) but the second is vastly different since the above result is the derivative of Eq. (2.2) w.r.t. r. Eq. (2.22) cannot yield all the asymptotics for $S^{r,1}_{p/q}(a)$ because we have already shown that Eq. (2.2) does not contain all the asymptotics for all values of p/q in $S^r_{p/q}(a)$.

The preceding analysis has relied on the fact that p is not equal to unity. When this is the case, the integral in Eq. (2.7) becomes

$$I = \frac{1}{2}\sum_{j=0}^{\infty} \left(1 + (-1)^j\right) \frac{\beta^j\, e^{i\pi j(2s+1/2)}}{a^{jq+q+\alpha+1}}\, \frac{\Gamma(jq + q + \alpha + 1)}{\Gamma(j+2)} \quad . \tag{2.23}$$

Effectively, this means that we must put $l = 0$ and $p = 1$ in our previous results. Hence, we find

$$\begin{aligned} S^r_{1/q}(a) &= \frac{q\Gamma(qr)}{a^{qr}} + \frac{1}{2}\,\delta_{r,1} + (2\pi)^{(1-q)/2}\,\sqrt{q}\,\sum_{n=1}^{\infty}\left(\frac{q}{a}\right)^{qr} \\ &\times \Bigg[\prod_{\Gamma}^{q,1}\left(1, r-1\right)\, {}_{q+1}\mathbf{F}_1\left(1, \Delta_{q-1}\left(q, qr+1\right); 2; \pm 2i\left(\frac{q}{a}\right)^q e^{2is\pi} n\pi\right)^+ \\ &\qquad - \left(r-1\right)\prod_{\Gamma}^{q,1}\left(1, r-1-\frac{1}{q}\right) \\ &\times {}_{q+1}\mathbf{F}_1\left(1, r, \Delta_{q-2}\left(q, qr+1\right); 2; \pm 2i\left(\frac{q}{a}\right)^q e^{2is\pi} n\pi\right)^+\Bigg] \quad , \end{aligned} \tag{2.24}$$

and

$$\begin{aligned} S_{1/q}(a) &= \frac{q\Gamma(q)}{a} + \frac{1}{2} + (2\pi)^{(1-q)/2}\,\frac{q^{q+1/2}}{a^q}\sum_{n=1}^{\infty}\prod_{\Gamma}^{q,1}\left(1, 0\right) \\ &\times {}_qF_0\left(1, \Delta_{q-2}\left(q, q+1\right); \pm ize^{2is\pi}\right)^+ \quad . \end{aligned} \tag{2.25}$$

It should be noted that the introduction of the hypergeometric functions in the above equations is a perfectly proper identification provided that their final utilisation as in Sec. 5.A is by way of their appropriate, convergent integral representation.

To complete this section, we utilise our result for the generalised Euler-Jacobi series to produce further extensions of it for special cases. First, if p/q is an integer, then Eq. (2.14) becomes

$$S_p\left(\frac{1}{\beta t}\right) = \sum_{n=0}^{\infty} e^{-n^p/(\beta t)} = (\beta t)^{1/p}\,\Gamma\left(\frac{1}{p+1}\right) + \frac{1}{2} + (2\pi)^{(p-1)/2}\,p^{-3/2}$$

$$\times\ (\beta t)^{1/p}\sum_{n=1}^{\infty}\sum_{l=0}^{\infty}(-1)^l\left(\frac{2\pi n\,(\beta t)^{1/p}}{p}\right)^{2l}\prod_{\Gamma}^{1,p}\left(\frac{2l+1}{p},0\right)$$
$$\times\ {}_1F_{p-1}\Big(1;\Delta_{p-2}\big(p,2l+2\big);\pm\,e^{i\pi p'/2}\left(\frac{2\pi n}{p}\right)^p\beta t\Big)^+ . \qquad (2.26)$$

Multiplying Eq. (2.26) by $(1-\beta)^{\mu-1}\beta^{\nu-1}$, setting s equal to zero and then integrating over β between 0 and 1, we get

$$\sum_{n=1}^{\infty}\Gamma(\mu)\left(\frac{n^p}{t}\right)^{\nu/2-1}e^{-n^p/2t}\,W_{1-\mu-\nu/2,(\nu-1)/2}\left(\frac{n^p}{t}\right)=\Gamma\left(\frac{1}{p+1}\right)\,B(\nu+1/p,\mu)$$
$$\times\ t^{1/p}-\frac{1}{2}\,B(\nu,\mu)\ +\ (2\pi)^{(p-1)/2}\,\frac{t^{1/p}}{p^{3/2}}\,\sum_{n=1}^{\infty}\sum_{l=0}^{p-1}(-1)^l\left(\frac{2\pi n\,t^{1/p}}{p}\right)^{2l}$$
$$\times\ \prod_{\Gamma}^{1,p}\left(\frac{2l+1}{p},0\right)B\big(\nu+(2l+1)/p,\mu\big)$$
$$\times\ {}_2F_p\Big(1,\nu+\frac{2l+1}{p};\nu+\mu+\frac{2l+1}{p},\Delta_{p-2}\big(p,2l+2\big);\pm\,z\Big)^+ , \qquad (2.27)$$

where $W_{\alpha,\beta}(z)$ is a Whittaker function, $B(\mu,\nu)$ is the beta function and $z=e^{i\pi p'/2}(2\pi n/p)^p t$. To obtain the above result, we have used Nos. 3.383(4) and 7.512(12) from Gradshteyn and Ryzhik [16]. The above result is only valid for Re $\mu>1$, Re $\nu>0$, Re $t>0$ and $p>2$. For the case of $\nu=2\mu+2$, Eq. (2.27) becomes

$$\sum_{n=1}^{\infty}\left(\frac{n^p}{t}\right)^{\mu+1/2}e^{-n^p/2t}K_{\mu+1/2}\left(\frac{n^p}{2t}\right)=\frac{\sqrt{\pi}}{p}\frac{\Gamma(2\mu+2+1/p)}{\Gamma(3\mu+2+1/p)}\,\Gamma(1/p)$$
$$-\ \frac{\sqrt{\pi}}{2}\frac{\Gamma(2\mu+2)}{\Gamma(3\mu+2)}+\frac{(2\pi)^{p/2}}{\sqrt{2}}\,\frac{t^{1/p}}{p^{3/2}}\sum_{n=1}^{\infty}\sum_{l=0}^{p-1}(-1)^l\left(\frac{2\pi n t^{1/p}}{p}\right)^{2l}$$
$$\times\ \prod_{\Gamma}^{1,p}\left(\frac{2l+1}{p},0\right)\ \frac{\Gamma(2\mu+2+(2l+1)/p)}{\Gamma(3\mu+2+(2l+1)/p)}$$
$$\times\ {}_2F_p\Big(1,2\mu+2+\frac{2l+1}{p};3\mu+2+\frac{2l+1}{p},\Delta_{p-2}\big(p,2l+2\big);\pm\,z\Big)^+ . \qquad (2.28)$$

If we return to Eq. (2.26), then using No. 3.471(9) from Ref. [16] we find

$$\int_0^{\infty}d\beta\,\beta^{s-1}e^{-\beta\alpha}S_p\left(\frac{1}{\beta t}\right)=\frac{\Gamma(s)}{\alpha^s}+2\sum_{n=1}^{\infty}\left(\frac{n^p}{\alpha t}\right)^{s/2}K_s\left(2\sqrt{\frac{n^p\alpha}{t}}\right) . \qquad (2.29)$$

Evaluating the appropriate integral for the r.h.s. of Eq. (2.26), we finally get for $p > 2$ and Re $s > 0$

$$\sum_{n=1}^{\infty} \left(\frac{n^p}{\alpha t}\right)^{s/2} K_s\left(2\sqrt{\frac{n^p\alpha}{t}}\right) = \frac{\Gamma(1/p)\,t^{1/p}}{2\,p\,\alpha^{s+1/p}}\,\Gamma(s+1/p) - \frac{\Gamma(s)}{4\alpha^s} + \frac{t^{1/p}}{2p^{3/2}}$$

$$\times\,(2\pi)^{(p-1)/2}\sum_{n=1}^{\infty}\sum_{l=0}^{p-1}(-1)^l\left(\frac{2\pi n\,t^{1/p}}{p}\right)^{2l}\frac{\Gamma(s+(2l+1)/p)}{\alpha^{s+(2l+1)/p}}\;\prod_{\Gamma}^{1,p}\left(\frac{2l+1}{p},0\right)$$

$$\times\;{}_2F_{p-1}\left(1, s+\frac{2l+1}{p}; \Delta_{p-2}\big(p,2l+2\big); \pm\left(\frac{2\pi n i}{p}\right)^p\frac{t}{\alpha}\right)^+, \tag{2.30}$$

where we have used No. 7.522(5) from Ref. [16]. An interesting feature of Eq. (2.30) is that if $2(\alpha/t)^{1/2}$ is replaced by a and s is put equal to $1/2$, then we obtain $S_{p/2}(a)$ as given by Eq. (2.14).

In this section we have presented exact results for both the series $S^r_{p/q}(a)$ and the generalised Euler-Jacobi series. These results could be expressed either in terms of hypergeometric functions as in Eqs. (2.13) and (2.14) or in terms of Meijer G-functions as in Eqs. (2.15) to (2.18). The latter forms, however, are only valid for $q+1 \le p$ or for $p = q$ and $|z| < 1$. Therefore, in Sec. 6 when we examine the asymptotic behaviour of the generalised Euler-Jacobi series for values of $p/q > 2$, we shall utilise both notations but when we examine the asymptotics of the series for $p/q < 2$ in Sec. 5, we can only utilise the hypergeometric version of the series.

3. PROPERTIES OF $S_{p/q}(a)$

To facilitate our examination of the asymptotic behaviour of the generalised Euler-Jacobi series in Sec. 6, we shall make some general observations about Eq. (2.14) in this section. Although we do not examine the asymptotic behaviour of $S^r_{p/q}(a)$ in great detail in this work, we shall, where possible, carry our observations about Eq. (2.14) to Eq. (2.13).

According to Luke [8], for $q \le p-2$ the complete asymptotic expansion of the hypergeometric function ${}_qF_{p-1}$ or its analogue Meijer G-function $G^{1,q}_{q,p}$ is given by

$${}_qF_{p-1}\left(\begin{matrix}\alpha_q\\ \rho_{p-1}\end{matrix}\,\middle|\,-z\right) = \frac{\Gamma(\rho_{p-1})}{\Gamma(\alpha_q)}\,G^{1,q}_{q,p}\left(-z\,\middle|\,\begin{matrix}1-\alpha_q\\ 1-\rho_{p-1}\end{matrix}\right) \sim \frac{\Gamma(\rho_{p-1})}{\Gamma(\alpha_q)}$$

$$\times\left[\sum_{k=0}^{r-1}\Gamma_p^{1,q}(k)\,K_{q,p-1}\big(ze^{-i\pi(2k+1)}\big) + \sum_{s=0}^{p-q-r-1}\bar{\Gamma}_p^{1,q}(s)\right.$$

$$\left.\times\;K_{q,p-1}\big(ze^{i\pi(2s+1)}\big) + L_{q,p-1}(z)\right], \tag{3.1}$$

where

$$K_{q,p-1}(z) = \frac{(2\pi)^{(q+1-p)/2}}{(p-q)^{1/2}} \, e^{(p-q)z^{1/(p-q)}} \, z^{\gamma} \left(1 + \sum_{r=1}^{\infty} N_r z^{-r/(p-q)}\right) \quad , \tag{3.2}$$

and

$$L_{q,p-1}(z) = \sum_{t=1}^{q} z^{-\alpha_t} \, \frac{\Gamma(\alpha_t)\,\Gamma(\alpha_q - \alpha_t)^*}{\Gamma(\rho_{p-1} - \alpha_t)} \times {}_pF_{q-1}\left(\begin{matrix} \alpha_t, 1 + \alpha_t - \rho_{q-1} \\ 1 + \alpha_t - \alpha_q^* \end{matrix} \middle| \frac{(-)^{p-q-1}}{z} \right) \quad . \tag{3.3}$$

In Eq. (3.1), $0 \le q \le p-2$ and r is an arbitrary integer such that $0 \le r \le p-q$ whilst $|\arg z| \le 2\pi - \delta$ with $\delta > 0$. The argument of z is also subject to $\delta_1 + (4r - 3p + 3q - 2)\pi/2 \le \arg z \le \delta_2 + (4r - p + q + 2)\pi/2$ where δ_1 and δ_2 are arbitrarily small. In the above $\Gamma(\rho_{p-1})$ and $\Gamma(\alpha_q)$ are respectively $p-1$ and q products of gamma functions containing $\rho_1, ..., \rho_{p-1}$ and $\alpha_1, ..., \alpha_q$. The same notation applies to $\Gamma(\alpha_q - \alpha_t)^*$ and $\Gamma(\rho_{q-1} - \alpha_t)$ except the variables in the gamma functions are $\alpha_1 - \alpha_t, ..., \alpha_q - \alpha_t$ and $\rho_1 - \alpha_t, ..., \rho_{q-1} - \alpha_t$ for each value of t. The asterisk means that $\Gamma(\alpha_q - \alpha_q)$ is excluded. In addition, $\Gamma_p^{1,q}(0) = \bar{\Gamma}_p^{1,q}(0) = 1$ whereas the other values are given on p. 196 of Ref. [8]. These will be evaluated when determining the asymptotics of the generalised Euler-Jacobi series in Secs. 5 and 6.

In Eq. (3.2),

$$\beta_0 \, \gamma = \frac{\beta_0 - 1}{2} + B_1 - C_1 \quad , \tag{3.4}$$

where $\beta_0 = p - q$, $B_1 = \sum_{t=1}^{q} \alpha_t$ and $C_1 = \sum_{t=1}^{p-1} \rho_t$. The N_r can be related to the c_r, which, in turn, can be evaluated by means of recursion relations originally proposed by Wright [18]. Thus

$$\beta_0^{-r} \, c_r \left(q, p \middle| \begin{matrix} \alpha_q \\ 1, \rho_{p-1} \end{matrix} \right) = N_r = N_r \left(q, p-1 \middle| \begin{matrix} \alpha_q \\ 1, \rho_{p-1} \end{matrix} \right) \quad , \quad N_0 = 1 \quad , \tag{3.5}$$

$$T_p(-r) - U_q(-r) = 0 \quad , \tag{3.6}$$

$$T_{p-1}(-r) - U_{q-1}(-r) = -\beta_0 \, r \quad , \tag{3.7}$$

and

$$\beta_0 \, r \, c_r = \sum_{s=1}^{p-1} T_{p-1-s}(s-r) \, c_{r-s} - \sum_{s=1}^{q-1} U_{q-1-s}(s-r) \, c_{r-s} \quad , \tag{3.8}$$

where the second sum is nil if $q \leq 1$ and $c_k = 0$ if $k < 0$. The T_s and U_s are evaluated by using

$$T(t) = \prod_{j=0}^{p-1} (t + \omega_j) \ , \tag{3.9}$$

and

$$U(t) = \prod_{j=1}^{q} (t + \lambda_j) \ , \tag{3.10}$$

which appear in the following equations:

$$T_s(-r) = \sum_{k=0}^{s} \frac{(-1)^{s-k}}{k!} \frac{T(k-r)}{(s-k)!} \ , \tag{3.11}$$

and

$$U_s(-r) = \sum_{k=0}^{s} \frac{(-1)^{s-k}}{k!} \frac{U(k-r)}{(s-k)!} \ . \tag{3.12}$$

In Eqs. (3.9) and (3.10),

$$\omega_j = \beta_0 \, \rho_j - \beta_0 + \beta_0 \, \gamma + 1 - \rho_0 \ , \tag{3.13}$$

and

$$\lambda_j = \beta_0 \, \alpha_j + \beta_0 \, \gamma + 1 - \rho_0 \ , \tag{3.14}$$

with $\rho_0 = 1$. The variable ρ_0 appears in the above equation because the recursion relations given by Eqs. (3.11) and (3.12) are obtained for a more general variant of $K_{q,p-1}$ in Luke [8]. When ρ_0 is set equal to unity, this result reduces to the result given by Eq. (3.2). The evaluation of the N_r will become clearer to the reader once we evaluate the asymptotics for the generalised Euler-Jacobi series for $p/q \geq 2$.

It should be noted that although Eq. (3.1) is an asymptotic approximation to ${}_qF_{p-1}$, it is actually derived from an exact result given as Theorem 3.1 in Sec. 5.9 of Luke [8]. According to this theorem, the Meijer G-function $G_{p,q}^{m,n}(z)$ can be expressed exactly in terms of three distinct sums; the first two involving $G_{p,q}^{q,0}(z \exp(i\pi(\nu - 2k)))$ and $G_{p,q}^{q,0}(z \exp(-i\pi(\nu - 2k)))$ separately and the last involving $G_{p,q}^{q,1}$. Provided the various conditions are obeyed, the last sum leads exactly to the $L_{q,p-1}$-term in Eq. (3.1). On the other hand, Luke shows in Sec. 5.7 that the dominant asymptotic terms of $G_{p,q}^{q,0}(z)$ can be written in terms of $H_{p,q}$, which eventually lead to the $K_{q,p-1}$-term in Eq. (3.1). Hence terms subdominant to the smallest exponential on the r.h.s. of Eq. (3.1) have been neglected. We shall see in Sec. 6.2 that our result for $S_3(a)$ can be expressed in terms of special functions, whose asymptotics are well-known. There we show also that the asymptotic form for $S_3(a)$ obtained from the asymptotics of the special functions is identical to the asymptotic form obtained via Eq. (3.1). Hence, the subdominant terms neglected in Eq. (3.1) are insignificant in the evaluation of the asymptotic forms for $S_{p/q}(a)$.

In this work we shall refer to the contribution emanating from the $K_{q,p-1}$-terms in Eq. (3.1) to the complete asymptotic series in a for the various series considered in the previous section as the K-asymptotics whereas those from the $L_{q,p-1}$-terms will be referred to as the L-asymptotics. From Eq. (3.3), we note that the L-asymptotics will produce a series in powers of a^q since $z = (q/a)^q(2n\pi/p)^p$. From Eq. (3.2) we note that the K-asymptotics will only consist of terms involving $\exp(s_k/a^{q/\beta_0})$ where the s_k have to be determined for each specific case. Therefore, the L-asymptotics must produce the power series in a as obtained by Berndt in Eqs. (2.2) and (2.3), whereas the K-asymptotics will yield the exponential terms missing from those equations.

An interesting feature of the K-asymptotics is that for each of the hypergeometric functions appearing in the generalised Euler-Jacobi series or for that matter in $S^r_{p/q}(a)$ there are terms which grow exponentially as $a \to 0$ since Re $s_k = \beta_0 \left(q^q(2n\pi/p)^p\right)^{1/\beta} \cos((2k+1)\pi/(2\beta_0))$ can be greater than zero. Since both series converge, this means that the positive Re s_k contribution from each of the hypergeometric functions must combine to vanish in a unique way. To see this more clearly, consider the summation over l in Eqs. (2.14) and (2.16) for p/q equal to 3, which yields after some simplification

$$\sum_{l=0}^{2}(-1)^l z^{2l/3}\, G^{1,1}_{1,3}\left(\pm z e^{i\pi/2}\,\middle|\, \begin{matrix}0\\ 0,-2l/3,(1-2l)/3\end{matrix}\right)^{+} = \Gamma\left(\frac{2}{3}\right)\, {}_0F_1\left(\frac{2}{3};\pm iz\right)^{+}$$

$$- z^{2/3}\,\Gamma\left(\frac{4}{3}\right)\Gamma\left(\frac{5}{3}\right)\, {}_1F_2\left(1;\frac{4}{3},\frac{5}{3};\pm iz\right)^{+} - iz^{1/3}\,\Gamma\left(\frac{4}{3}\right)\, {}_0F_1\left(\frac{4}{3};\pm iz\right)^{-}. \quad (3.15)$$

As mentioned earlier, it will be shown in Sec. 6 that the above result can be expressed in known functions. Since $\beta_0 = 2$ for all the hypergeometric functions in Eqs. (3.7) and (3.8), Re $s_k = 2(2n\pi/3)^{3/2}\cos((2k+1)\pi/4)$, which is positive for $k = 0$. Hence the leading order terms from the K-asymptotics for each of the hypergeometric functions in Eq. (3.15) diverge as $a \to 0$ but since Eq. (2.14) is convergent, these must cancel. However, because each of the hypergeometric functions has different parameters, Eqs. (3.4) to (3.12) would seem to indicate that the recursion relations for obtaining the N_r are different. This is obviously not the case or else our result would be divergent. In actual fact we shall show in Sec. 6 that the first and third hypergeometric functions in Eq. (3.15) will yield the same recursion relation for N_r when evaluating the T_s as given by Eqs. (3.9) and (3.11). It will also be seen that the recursion relation will be different for the ${}_1F_2$ since the T-function from Eq. (3.9) is different. However, after some algebraic manipulation, we will show that the recursion relation is in fact identical to the recursion relation obtained for the ${}_0F_1$ hypergeometric functions.

As we shall see in Sec. 6, the same behaviour regarding the recursion relation applies to other integer cases. From this, we deduce that there are groups or families of hypergeometric functions, which possess similar K-asymptotics where the differences occur in the power of z multiplying the asymptotic forms, herein given as z^γ, and in the phase of the trigonometric functions arising from the complex parts of the exponentials in Eq. (3.2). We shall also see that when the z^γ for a hypergeometric function or a Meijer G-function is multiplied by its $z^{2l/p}$ in the l-summation in Eqs. (2.14) and (2.16), all algebraic powers in $S_p(a)$ are identical, which means that they can be combined. Then we shall see that the differences in the phases of the trigonometric functions are required to enable the complete cancellation of all exponentially growing terms arising from the K-asymptotics, thereby yielding a convergent asymptotic result for $S_p(a)$. This, therefore, means that the dominant terms from the K-asymptotics of the hypergeometric functions in the generalised Euler-Jacobi series or for that matter $S^r_{p/q}(a)$ are redundant. As a consequence, we will need to present a detailed evaluation of the asymptotics for each hypergeometric function whereby subdominant terms are retained in order to obtain a convergent asymptotic result for $S_{p/q}(a)$. These evaluations are, of course, exhibited in Secs. 5 and 6.

Because of the arbitrariness of r in the summation over k and s in Eq. (3.1), different asymptotic expansions for the same hypergeometric function can be obtained for a common region of the complex plane. For example, Luke gives the following asymptotic expansions for ${}_pF_p\left(\begin{matrix}\alpha_p\\ \rho_p\end{matrix}\middle| z\right)$ as $z \to \infty$:

$$ {}_pF_p\left(\begin{matrix}\alpha_p\\ \rho_p\end{matrix}\middle| z\right) \sim \frac{\Gamma(\rho_p)}{\Gamma(\alpha_p)} \left[K_{p,p}(z) + L_{p,p}(ze^{-i\pi})\right] , \tag{3.16} $$

where $\delta - \pi/2 \le \arg z \le 3\pi/2 - \delta$, $\delta > 0$ (small) and

$$ {}_pF_p\left(\begin{matrix}\alpha_p\\ \rho_p\end{matrix}\middle| z\right) \sim \frac{\Gamma(\rho_p)}{\Gamma(\alpha_p)} \left[K_{p,p}(z) + L_{p,p}(ze^{i\pi})\right] , \tag{3.17} $$

where $\delta - 3\pi/2 \le \arg z \le \pi/2 - \delta$. As Luke points out, the apparent discrepancy for $|\arg z| < \pi/2$ is attributed to the Stokes phenomenon, which is concerned with the behaviour of small exponentials masked by large ones in asymptotic approximations. Once an anti-Stokes line is crossed, the small exponential becomes the dominant one while the previously dominant one becomes the small exponential. For a complete exposition of this phenomenon, the reader is directed to Morse and Feshbach [19], Dingle [3] and to the interesting article by Berry [20]. In particular, Berry has shown how the changes in the multipliers of the exponentials of asymptotic expansions across anti-Stokes lines are not discontinuous but are smooth provided the asymptotic expansion of the dominant exponential is truncated near its least term. In Secs. 5 and 6, we shall i indicate to the reader how the Stokes phenomenon affects our analysis of the $p/q = 4/3$, $3/2$ and 3 cases but then will appeal to the reality and convergence of the general Euler-Jacobi series to resolve the discrepancies produced by the phenomenon.

We also show that for even integers, i.e. $p/q = 2k$ where $k \in \mathbf{Z}^+$, our result for the generalised Euler-Jacobi series does not produce a series in powers of a whereas for $p/q = 2k+1$, our result does produce a series in powers of a as obtained by Ramanujan and Berndt (see Eq. (2.4)). Our answer for the generalised Euler-Jacobi series (Eq. (2.14)) contains the following expression:

$$
{}_{\mathbf{q}}\mathbf{F}_{\mathbf{p-1}}\left(\frac{2l+1}{p}, 0, \pm z e^{i\pi p'/2}\right)^{+} = \prod_{\Gamma}^{q-1,p-1}\left(\frac{2l+1}{p}, 0\right)
\times {}_{q}F_{p-1}\Big(1, \Delta_{q-2}(q, \tau(1,0))\,; \Delta_{p-2}(p, 2l+2)\,; \pm z e^{i\pi p/2}\Big)^{+} . \tag{3.18}
$$

By choosing an appropriate value of s, and putting $p = 2k$ and $q = 1$, we find that the above result becomes

$$
{}_{1}\mathbf{F}_{\mathbf{2k-1}}\left(\frac{2l+1}{2k}, 0, \pm z e^{i\pi k}\right)^{+} = \prod_{m=0}^{2k-2}\Gamma\left(\frac{2l+m+2}{2k}\right)^{-1}
\times {}_{1}F_{2k-1}\left(1; \frac{2k+2l-\{j\}}{2k}; \pm\frac{e^{ik\pi}}{a}\left(\frac{\pi n}{k}\right)^{2k}\right)^{+} , \tag{3.19}
$$

where the sum over l ranges from 0 to $2k-1$ and the notation $(2k+2l-\{j\})/(2k)$ means that each of the values from the set $\{j\} = \{0, 1, \cdots, 2k-2\}$ must be included in the parameters for the hypergeometric functions. Since j is a maximum of $2k-2$, there will always be one value of l less than $k-1$ such that $(2k+2l-j)/(2k)$ is equal to 1. Then ${}_1F_{2k-1}$ hypergeometric functions in Eq. (3.19) reduce to ${}_0F_{2k-2}$ hypergeometric functions. When this occurs, the L-asymptotics for these hypergeometric functions become $L_{0,2k-2}(\pm e^{ik\pi}\,(n\pi/k)^{2k}\,a)$, which vanishes since the summation in Eq. (3.3) does not exist. For $l > k-1$, there will always be a value of $\{j\}$ such that $(2k+2l-j)/2k = 2$; e.g. for $l = 2k-1$, $(2k+2l-j)/2k = 2$ when $j = 2k-2$. Hence for $k-1 \le l \le 2k-1$, Eq. (3.19) becomes

$$
{}_{1}\mathbf{F}_{\mathbf{2k-1}}\left(\frac{2l+1}{2k}, 0, \pm z e^{i\pi k}\right)^{+} = \left[\prod_{m=0}^{2k-2}\Gamma\left(\frac{2l+m+2}{2k}\right)\right]^{-1}
\times {}_{1}F_{2k-1}\left(1; 2, \frac{2k+2l-\{j\}'}{2k}; \pm\frac{e^{ik\pi}}{a}\left(\frac{\pi n}{k}\right)^{2k}\right) , \tag{3.20}
$$

where $\{j\}'$ excludes the value of j equal to $2l-2k$. If we now consider the series expansion of the ${}_1F_{2k-1}$ hypergeometric functions in Eq. (3.20), the gamma function produced by the upper index of unity will cancel the one appearing in the denominator, thus leaving us with a series of the following form:

$$\left[\prod_{m=0}^{2k-2} \Gamma\left(\frac{2l+m+2}{2k}\right)\right]^{-1} {}_1F_{2k-1}\left(1;2,\frac{2k+2l-\{j\}'}{2k};\pm z\right) = \sum_{n=0}^{\infty} \frac{(\pm z)^n}{\Gamma(n+2)\,\Gamma(n+(2k+2l-\{j\}')/2k)} \quad . \tag{3.21}$$

As we shall see when discussing the specific cases in Secs. 5 and 6, the sum of the above series, i.e. ${}_1F_{2k-1}(1;2,(2k+2l-\{j\}')/2k;\pm z)^+$, can be converted to ${}_0F_{2k-2}(1,(2k+2l-\{j\}')/2k-1;\pm z)^-$ by making the substitution $n'=n+1$. Hence, all the hypergeometric functions for $p/q=2k$ reduce to ${}_0F_{2k-2}$ hypergeometric functions, whose L-asymptotics vanish. This means, therefore, that there can be no power series in a for $p/q=2k$ since the K-asymptotics only yield exponentials as discussed previously.

The situation, however, is different when $p/q=2k+1$. Then Eq. (3.18) becomes

$$\mathbf{{}_1F_{2k}}\left(\frac{2l+1}{2k+1},0,\pm e^{i\pi(k+1/2)}\right)^+ = \left[\prod_{m=0}^{2k-1} \Gamma\left(\frac{2l+m+2}{2k+1}\right)\right]^{-1} \times {}_1F_{2k}\left(1;\frac{2l+2}{2k+1},\frac{2l+3}{2k+1},\ldots,\frac{2l+2k+1}{2k+1};\pm\frac{1}{a}\left(\frac{2n\pi i}{2k+1}\right)^{2k+1}\right) \quad , \tag{3.22}$$

where l ranges from 0 to $2k$. We note that when $l=k$, the indices for the hypergeometric functions in Eq. (3.22) range from $1+(2k+1)^{-1}$ to $1+2k/(2k+1)$. That is none of the indices equals 1 or 2. Hence for $l=k$ the ${}_1F_{2k}$ hypergeometric functions cannot reduce to ${}_0F_{2k-1}$ hypergeometric functions. This means that the asymptotic expansion of the ${}_1F_{2k}$ hypergeometric functions will consist of both L- and K-asymptotics. For all other values of k the ${}_1F_{2k}$ hypergeometric functions in Eq. (3.22) will reduce to ${}_0F_{2k-1}$ hypergeometric functions similar to the situation encountered when $p/q=2k$.

Thus for $p=2k+1$ and $q=1$, we have found that the $l=k$ term in the l-summation of Eq. (2.14) is the only term that has L-asymptotics or a power-series in a. Extracting this term from the result for $S_{2k+1}(a)$, we get

$$\frac{(-1)^k}{a}\,\frac{(2\pi)^k}{(2k+1)^{3/2}} \sum_{n=1}^{\infty} \left(\frac{2\pi n}{2k+1}\right)^{2k} \left[\prod_{m=0}^{2k} \Gamma\left(\frac{2k+m+2}{2k+1}\right)\right]^{-1} \times {}_1F_{2k}\left(1;\frac{2k+2}{2k+1},\ldots,\frac{4k+1}{2k+1};\pm\frac{1}{a}\left(\frac{2n\pi i}{2k+1}\right)^{2k+1}\right)^+ \quad . \tag{3.23}$$

The L-asymptotics for $S_{2k+1}(a)$, therefore, can be written as

$$S^L_{2k+1}(a) = \frac{(-1)^k}{a}\,\frac{(2\pi)^k}{(2k+1)^{3/2}} \sum_{n=1}^{\infty} \left(\frac{2\pi n}{2k+1}\right)^{2k} L_{1,2k}\left(\pm z e^{i\pi/2}\right)^+ \quad , \tag{3.24}$$

where $z = (2n\pi/(2k+1))^{2k+1}a^{-1}$. Now from Eq. (3.3), we find that

$$L_{1,2k}\left(\pm ze^{i\pi/2}\right)^{+} = i\left[z\prod_{l=2}^{2k+1}\Gamma\left(\frac{2k+l}{2k+1}-1\right)\right]^{-1}$$
$$\times\ {}_{2k+1}F_0\left(1,\frac{2k+2}{2k+1}-1,\cdots,\frac{4k+1}{2k+1}-1;\mp iz^{-1}\right)^{-} . \tag{3.25}$$

Application of Gauss' multiplication formula (No. 8.335 of Ref. 5) to the series expansion of the above hypergeometric functions yields

$$L_{1,2k}\left(\pm ze^{i\pi/2}\right)^{+} = 2\left[z\prod_{l=2}^{2k+1}\Gamma\left(\frac{2k+l}{2k+1}-1\right)\right]^{-1}$$
$$\times\sum_{m=0}^{\infty}(-1)^m\,\frac{z^{-(2m+1)}}{\Gamma(2m+2)}\,\frac{\Gamma\Big((2k+1)(2m+1)+1\Big)}{(2k+1)^{(2m+1)(2k+1)}} . \tag{3.26}$$

Introducing the above result into Eq. (3.24) and interchanging the order of the summations, we get

$$S^L_{2k+1}(a) = 2\sum_{m=0}^{\infty}(-1)^{k+m}\,\frac{\Gamma\Big((2k+1)(2m+1)+1\Big)}{(2\pi)^{(2m+1)(2k+1)+1}}$$
$$\times\ \frac{a^{2m+1}}{\Gamma(2m+2)}\,\zeta\Big((2m+1)(2k+1)+1\Big) , \tag{3.27}$$

where once again we have utilised Gauss' multiplication formula. We have also identified the n-summation as a Riemann zeta function. Now if we let $N\to\infty$ and put $r=1$ with $p/q=2k+1$ in Eq. (2.2) and disregard the $l=0$ term and $q\Gamma(qr/p)/(pa^{qr/p})$, which account for the terms outside the sum on the r.h.s. of Eq. (2.14), then

$$I'_{\infty,\infty} = \sum_{m=0}^{\infty}(-1)^{2m+1}\,\frac{a^{2m+1}}{(2m+1)!}\,\zeta\Big(-(2k+1)(2m+1)\Big) , \tag{3.28}$$

where the prime denotes the exclusion of some terms. Now using the Riemann recursion relationship (No. 9.535 of Gradshteyn and Ryzhik [16]), we finally get Eq. (3.27). Thus the L-asymptotics for p/q equal to $(2k+1)$ produce the result obtained by Ramanujan and Berndt for the generalised Euler-Jacobi series. As a result of the preceding discussion, we can see immediately that the ${}_1F_2$ hypergeometric function appearing in Eq. (3.15) will produce the 'zeta' series as given by Eq. (3.27) for $S_3(a)$.

The final point, that we discuss in this section, is whether it is possible not to obtain a zeta series for values of q other than unity, assuming that p/q has been completely factorised. For the ${}_qF_{p-1}$ hypergeometric functions to reduce to ${}_0F_{p-q-1}$, we note that $p-q-1\geq 0$, i.e. $p\geq q+1$; otherwise, we must obtain a contribution from the L-asymptotics. To be certain that there is no L-asymptotic contribution when $p \geq q+1$, we must have $\{(2l+1)/p+1/q, (2l+1)/p+2/q, \cdots, (2l+1)/p+(q-1)/q\} \subset \{(2l+2)/p+j, (2l+3)/p+j, \cdots, (2l+p)/p+j\}$ where j is an arbitrary integer. In our examination of the $p/q = 2k$ case, this integer was equal to 0 or 1 depending on the value of l. For the first set to be contained in the second, there must be some positive integer m such that $(2l+1)/p+1/q = (2l+2+m)/p+k$ or $p/q = m+1+kp$. Since $m+1+kp$ is an integer and $q \neq 1$, this implies p/q must have a common factor of q, which contradicts our earlier condition. Hence for all values of $q > 1$, we must obtain a contribution from the L-asymptotics, which will be observed when we consider p/q equal to 4/3 and 3/2 in Sec. 5.

4. STEEPEST DESCENT

In this section we investigate the utility of applying the method of steepest descent to the integral in Eq. (2.7). We shall demonstrate for the generalised Euler-Jacobi series that problems arise when considering values of $p/q > 4$. Nevertheless, we shall evaluate the leading terms of the asymptotic expansions for p/q equal to 3 and 4. This section will serve as a useful check on the results obtained in Sec. 6.

The integral in Eq. (2.7) can be transformed into the following contour integral:

$$I = \mathrm{Im}\left\{\frac{p}{q}\int_C dz\; a^{p(1+\alpha)/q(1-p/q)}\, z^{(\alpha+1)p/q-1} \exp\Big(-a^{1/(1-p/q)}\, f(z)\Big)\right\} \quad , \tag{4.1}$$

where $f(z) = z^{p/q} - i\beta z$, $\beta = 2n\pi$ as in Sec. 2 and C is the line contour along the real axis. The principal value for non-integer powers of z is fixed by having the branch line emanate from $z = 0$ to infinity along a ray lying in the domain, Re $z < 0$, e.g. the negative real axis. For $a \ll 1$, we see immediately from Eq. (4.1) that the method of steepest descent can only be applied if $p > q$. For $p < q$, the method is only valid if $a \gg 1$, which, of course, is not the concern of this work.

Saddle points occur when $f'(z) = 0$, which yields

$$z_0 = \left(\frac{\beta\, q\, e^{i\pi(2k+1/2)}}{p}\right)^{1/(p/q-1)} \quad , \tag{4.2}$$

where k is an arbitrary integer. Here we encounter the first problem in applying the method of steepest descent to Eq. (4.1); which saddle point do we choose? Murray [21] states that the chosen saddle point must allow the contour C to be deformed into a path of steepest descent through it. Hence, we should consider $|\arg z_0| < \pi/2$.

The paths of steepest ascent and descent are both defined by the same equation, Im $f(z) = \text{Im } f(z_0)$. This equation is

$$r^{p/q}\sin(p\,\theta/q) + \beta r\cos(\theta) = \left(\frac{\beta q}{p}\right)^{p/(p-q)}$$
$$\times \left[\sin\left(\frac{\pi(2k+1/2)p}{p-q}\right) + \frac{p}{q}\cos\left(\frac{\pi(2k+1/2)p}{p-q}\right)\right] \ , \tag{4.3}$$

where $z = re^{i\theta}$. Eq. (4.3) is a difficult equation to solve but it does show that $r = 0$ is not a solution. Hence, the origin is not on the path of steepest descent. We shall see, shortly, that this is important in determining the asymptotics for $S_3(a)$.

In view of the complexity of Eq. (4.3) and the difficulty in choosing the appropriate saddle points let us consider p/q equal to 3. Then there is only one appropriate saddle point, $z_0 = (\beta/3)e^{i\pi/4}$. In Cartesian co-ordinates Eq. (4.3) becomes

$$\beta x - 3yx^2 + y^3 = \sqrt{2}(\beta/3)^{3/2} \ . \tag{4.4}$$

When $x = 0$ then $y = 2^{1/6}(\beta/3)^{1/2}$, which supports our earlier statement concerning the origin not being on the path of the steepest descent. The solution to Eq. (4.4) yields both the paths of steepest ascent and descent for $S_3^r(a)$. After some analysis, we find that the path of steepest descent is given by

$$x = \beta/6y - (y^2/3 + \beta^2/36y^2 - \sqrt{2}(\beta/3)^{3/2}/3y)^{1/2} \ . \tag{4.5}$$

Thus, as $y \to \infty$, $x \to -y/\sqrt{3}$, i.e. $\arctan(y/x) \to -\pi/3$.

We can now deform the contour along the real axis to the path of steepest descent. However, we need to make one modification. To ensure that the contribution from the arc integral vanishes in the limit as $R \to \infty$, we deform the path of steepest descent so that as $y \to \infty$, $\arctan(y/x) \to -\pi/6$. This should have no effect on the steepest descent method since the dominant contribution comes from the vicinity of the saddle point. As a consequence, the contour integral in Eq. (4.1) can be approximated by

$$\int_C dz\ z^{3\alpha+2}e^{-f(z)/\sqrt{a}} \approx \int_0^{2^{1/6}(\beta/3)^{1/2}} dt\ t^{3\alpha+2}\ e^{i\pi(3\alpha+3)/2}$$
$$\times\ \exp\Big((-\beta t + it^3)/\sqrt{a}\Big) + \int_{-\infty}^{\infty} dt\ z_0^{3\alpha+2}\exp\Big((f(z_0) - t^2)/\sqrt{a}\Big)\left(\frac{dz}{dt}\right) \ , \tag{4.6}$$

where the first integral arises because the steepest descent path intercepts the imaginary axis and the second integral is the contribution from near the saddle point along the path of steepest descent. As we shall see shortly, the first integral will be used to derive the 'zeta' series while the second will yield the oscillatory, exponentially decaying terms missing from the Ramanujan-Berndt result given by Eqs. (2.2) and (2.3).

If we denote the first integral in Eq. (4.6) by I_1, then making the change of variable $y = a^{-1/2}t$ allows us to write it as

$$I_1 \overset{a\to 0}{\approx} \left(ae^{i\pi}\right)^{(3\alpha+3)/2} \int_0^{2^{1/6}(\beta/3)^{1/2}a^{-1/2}} dy\, e^{-\beta y} y^{3\alpha+2} \sum_{m=0}^{\infty} \frac{(iay^3)^m}{m!} \,, \tag{4.7}$$

where we have expanded the oscillatory exponential since a is small. Interchanging the order of the summation and integral by appealing to absolute convergence and then utilising No. 3.381 from Gradshteyn and Ryzhik [16], we obtain

$$I_1 = (ae^{i\pi})^{(3\alpha+3)/2} \sum_{m=0}^{\infty} \frac{(ae^{i\pi/2})^m}{m!} \beta^{-(3m+3\alpha+3)}$$
$$\times\ \gamma\Big(3m+3\alpha+3, 2^{1/6}(\beta/3)^{1/2}\beta a^{-1/2}\Big) \overset{a\to 0}{\approx} (ae^{i\pi})^{(3\alpha+3)/2}\beta^{-(3\alpha+3)}$$
$$\times \sum_{m=0}^{\infty} \frac{\Gamma(3m+3\alpha+3)}{\Gamma(m+1)} \left(\frac{ae^{i\pi/2}}{\beta^3}\right)^m \,, \tag{4.8}$$

where $\gamma(a,b)$ is an incomplete gamma function and $\text{Re}\,(3m+3\alpha+3) > 0$.

At the saddle point, $f(z_0) = 2(\beta/3)^{3/2}e^{3i\pi/4}$ and $dz/dt = (3\beta)^{-1/4}e^{-i\pi/8}$. If we denote the second integral in Eq. (4.6) by I_2, then we find

$$I_2 = \sqrt{\frac{\pi}{3}} \left(\frac{\beta}{3}\right)^{3(\alpha+1)/4} a^{1/4} \exp\left(-2\left(\frac{\beta}{3}\right)^{3/2} \frac{e^{-i\pi/4}}{\sqrt{a}} + \frac{3i\pi(2\alpha+1)}{8}\right) . \tag{4.9}$$

Combining Eqs. (4.8) and (4.9), we get

$$I \overset{a\to 0}{\approx} 3\sum_{m=0}^{\infty} \frac{\Gamma\big(3(m+\alpha+1)\big)}{\Gamma(m+1)} \frac{a^m}{\beta^{3(m+\alpha+1)}} \sin\left(\frac{(3\alpha+m+3)\pi}{2}\right)$$
$$+\ \frac{3\sqrt{3\pi}}{a^{3\alpha/2+5/4}} \left(\frac{\beta}{3}\right)^{3(\alpha+1)/4} e^{-\kappa} \sin\left(\kappa + \frac{3\pi(\alpha+1)}{8}\right) \,, \tag{4.10}$$

where $\kappa = (2\beta/3a)^{1/2}\beta/3$. Finally, substituting Eq. (4.10) into Eq. (2.6) gives

$$S_3^r(a) \overset{a\to 0}{\approx} \frac{\Gamma(r/3)}{3a^{r/3}} + \frac{1}{2}\,\delta_{r,1} + \frac{3}{\pi}\sum_{n=1}^{\infty}\frac{1}{n}\left[\sum_{m=0}^{\infty}\frac{\Gamma(3m+r+2)}{\Gamma(m+1)}\right.$$
$$\times\ \frac{a^{m+1}}{\beta^{3m+r+2}} \sin\left(\frac{(m+r+2)\pi}{2}\right) + \sqrt{\frac{\pi}{3}}\left(\frac{\beta}{3}\right)^{(r+2)/4} a^{1/4-r/2}$$
$$\times\ e^{-\kappa}\cos\left(\frac{\kappa+\pi(r-2)}{8}\right) - \frac{(r-1)}{3\beta^{r-3}}\sum_{m=0}^{\infty}\frac{\Gamma(3m+r-1)}{\Gamma(m+1)}\left(\frac{a}{\beta^3}\right)^m$$

$$\times\ \sin\left(\frac{(r-1+m)\pi}{2}\right) - \sqrt{\frac{\pi}{3}}\,\frac{(r-1)}{3}\left(\frac{\beta}{3}\right)^{(r-1)/4} a^{3/4-r/2}$$
$$\times\, e^{-\kappa}\sin\left(\kappa + \frac{\pi(r-1)}{8}\right)\Bigg] \ , \tag{4.11}$$

with $\beta = 2n\pi$ as stated previously. As an aside, $S_3^{r,k}(a)$ can be evaluated by taking the k-fold derivative of Eq. (4.11) w.r.t. r as we mentioned in Sec. 2. To facilitate this calculation, each derivative of the gamma function should be replaced by $\Gamma(r+c)\Psi(r+c)$ where c equals either $3m+2$ or $3m-1$. Furthermore, only the leading term must be retained since we have only evaluated the leading term in Eq. (4.11).

If we put $r = 1$ in Eq. (4.11), then we obtain the result for the generalised Euler-Jacobi series for $p/q = 3$, which is

$$S_3(a) \overset{a\to 0}{\approx} \frac{\Gamma(1/3)}{3\,a^{1/3}} + \frac{1}{2} + 2\sum_{n=1}^{\infty} \frac{\sqrt{\pi}}{(6n\pi a)^{1/4}}\, e^{-\kappa}\,\cos(\kappa - \pi/8)$$
$$- \frac{a}{8\pi^4}\sum_{n=0}^{\infty}(-1)^n\left(\frac{a}{8\pi^3}\right)^{2n}\frac{\Gamma(6n+4)}{\Gamma(2n+2)}\,\zeta(6n+4)\ . \tag{4.12}$$

To obtain the last term, we have interchanged the two summations and replaced the dummy index m by n. We note that Eq. (4.12) corresponds to the $k = 1$ result for Eq. (3.27), which we have noted is the series obtained by Ramanujan and Berndt. Thus the other summation corresponds to the oscillatory exponential term missing in their result. However, as we shall see in Sec. 6, the series involving the exponential in Eq. (4.12) is only the leading term of a double series involving oscillatory exponentials.

For higher values of p/q, determining the path of steepest descent and choosing the appropriate saddle point become increasingly difficult. Although determining the saddle point for the case of p/q equal to 4 for the generalised Euler-Jacobi series is not difficult, determination of the path of steepest descent is. For $p/q = 4$, we find from Eq. (2.6) that

$$S_4(a) = \frac{\Gamma(1/4)}{4a^{1/4}} + \frac{1}{2} + 2\ \mathrm{Re}\sum_{n=1}^{\infty}\int_C dz\ a^{-1/3} e^{2in\pi z a^{-1/3} - z^4 a^{-1/3}}\ , \tag{4.13}$$

where C is once again along the positive real axis. From the above equation the saddle point, $(2n\pi/4)^{1/3}e^{i\pi/6}$, can be chosen. Then the equation for both the paths of steepest ascent and descent is

$$2n\pi x - 4yx^3 + 4y^3x = 3(2n\pi/4)^{4/3}\sin(2\pi/3)\ . \tag{4.14}$$

This indicates that the intercept is now $x = (3/4)(2n\pi/4)^{1/3}\sin(2\pi/3)$.

Although resolving the paths of steepest ascent and descent is a formidable task requiring the solution of Eq. (4.14), we can use a heuristic approach to determine $S_4(a)$ as $a \to 0$. First, we note from the previous section that for $p/q = 4$ there will be no zeta series. Furthermore, since the oscillatory exponential for $S_3(a)$ comes from the vicinity of the saddle point, the integral in Eq. (4.13) near this point becomes

$$\int_C dz\, a^{-1/3} \exp\bigl(2in\pi z a^{-1/3} - z^4 a^{-1/3}\bigr) \overset{a\to 0}{\approx} \sqrt{\frac{\pi}{6}} \left(\frac{2}{n\pi\sqrt{a}}\right)^{1/3} \times \exp\Bigl(-\frac{3\alpha_0}{2}\,(1+\sqrt{3}\,i) - \frac{i\pi}{6}\Bigr) \; , \tag{4.15}$$

where $\alpha_0 = (n\pi/2a^{1/4})^{4/3}$. Introducing this into Eq. (4.15) yields

$$S_4(a) \overset{a\to 0}{\approx} \frac{\Gamma(1/4)}{4a^{1/4}} + \frac{1}{2} + \sum_{n=1}^{\infty} \frac{2^{5/6}\,\pi^{1/6}}{\sqrt{3}\,n^{1/3}\,a^{1/6}}\, e^{-3\alpha_0/2} \times \cos\Bigl(3\sqrt{3}\,\alpha_0/2 - \pi/6\Bigr) \; . \tag{4.16}$$

As expected, this approach only yields the dominant leading order term. Although the higher order terms can be obtained after much work (see for example Ch. VI of Ref. [3]), we shall show in Sec. 6 how the entire series follows naturally from Eq. (2.14). There we shall also show that as p/q increases, additional oscillatory exponential series begin to emerge, which are only remotely accessible, if at all, by using the method of steepest descent.

5. SPECIAL CASES OF $S_{p/q}(a)$ FOR $p/q < 2$

In this section we consider numerous cases of the generalised Euler-Jacobi series for p/q less than 2. Since the techniques used to determine the asymptotics for $p/q < 1$ are different from those employed for $p/q > 1$, we divide this section into three subsections.

A. $p/q < 1$

The three cases considered here are p/q equal to 1/3, 1/2 and 2/3. For $p/q = 1/3$, Eq. (2.14) becomes

$$S_{1/3}(a) = \frac{3\Gamma(3)}{a^3} + \frac{1}{2} + \frac{27\sqrt{3}}{2\pi a^3}\,\Gamma(4/3)\Gamma(5/3) \times \sum_{n=1}^{\infty} {}_3F_0\Bigl(1, 4/3, 5/3; \pm e^{i\pi/2} z\Bigr)^{+} \; , \tag{5.1}$$

where $z = 2n\pi(3/a)^3$. The asymptotic analysis given in Sec. 3 cannot be applied to the hypergeometric function in the above equation and furthermore, a representation for it does not appear in Prudnikov et al [22]. In fact, for all $p/q < 1$, the series representation for the hypergeometric functions comprising the generalised Euler-Jacobi inversion formula is divergent, which is consistent with Eq. (2.8). However, as mentioned just below this equation, we shall use the Borel summation technique to recast the various divergent series encountered in this section as convergent integrals, which can then be evaluated by various techniques. Thus, by introducing the series expansion for the ${}_3F_0$ and utilising the multiplication formula for the gamma function, we find that Eq. (5.1) can be written as

$$S_{1/3}(a) = \frac{3\Gamma(3)}{a^3} + \frac{1}{2} + \frac{3}{a^3} \sum_{n=1}^{\infty} \sum_{k=0}^{\infty} \left(\frac{z}{27}\right)^k \frac{\Gamma(3k+3)}{\Gamma(k+1)} \left(e^{i\pi k/2} + e^{-i\pi k/2}\right) . \quad (5.2)$$

Now introducing the integral representation for $\Gamma(3k+3)$ yields

$$S_{1/3}(a) = \frac{3\Gamma(3)}{a^3} + \frac{1}{2} + \frac{3}{a^3} \sum_{n=1}^{\infty} \frac{18}{z} \int_0^{\infty} dt\, e^{-t} \sin\left(zt^3/27\right) . \quad (5.3)$$

The integral in Eq. (5.3) does not appear in the appropriate sections of tables of integrals, Refs. [16,23,24]; so we evaluate it here by contour integration.

The integral in Eq. (5.3) can be written as

$$I = \frac{1}{2i} \int_C dt\, e^{-as} \left(e^{ibs^3} - e^{-ibs^3}\right) , \quad (5.4)$$

where C is the positive real axis, Re $a > 0$ and $b > 0$. To evaluate the first integral in Eq. (5.4), we deform the contour into the arc $s = Re^{i\theta}$ where θ ranges from 0 to $\pi/4$ and the ray arg $s = \pi/4$. Then Jordan's inequality can be used to show that the arc integral vanishes as $R \to \infty$. The second integral in Eq. (5.4) is evaluated by following the same procedure except that the arc ranges from 0 to $-\pi/4$ and the ray is now arg $s = -\pi/4$. Hence, we find that Eq. (5.4) becomes

$$I = \frac{1}{2i} \int_0^{\infty} dt\, \left[e^{i\pi/4} e^{-e^{i\pi/4}(at+bt^3)} - e^{-i\pi/4} e^{-e^{-i\pi/4}(at+bt^3)}\right] . \quad (5.5)$$

Putting $a = 3b\beta^2/4$ and making the substitution $t = \beta \sinh y$, we find after some manipulation that

$$I = \frac{\beta}{2i} \int_0^{\infty} dy\, \cosh y \left[\exp\left(i\pi/4 - e^{i\pi/4}(b\beta^3/4) \sinh 3y\right) - \exp\left(-i\pi/4 - e^{-i\pi/4}(b\beta^3/4) \sinh 3y\right)\right] . \quad (5.6)$$

Making the substitution $t = 3y$ and noting that $\mathrm{Re}\,(e^{\pm i\pi/4}(b\beta^3/4)) > 0$, so that we can use No. 3.547(3) from Gradshteyn and Ryzhik [16], we finally get

$$S_{1/3}(a) = \frac{3\Gamma(3)}{a^3} + \frac{1}{2} + \frac{27\pi}{ia^3} \sum_{n=1}^{\infty} \frac{1}{z^{3/2}} \left[e^{i\pi/4} \left\{ \tan\frac{\pi}{6} \left[\mathbf{J}_{1/3}\left(\frac{2e^{i\pi/4}}{\sqrt{z}}\right) \right.\right.\right.$$
$$\left. - J_{1/3}\left(\frac{2e^{i\pi/4}}{\sqrt{z}}\right)\right] - \left[\mathbf{E}_{1/3}\left(\frac{2e^{i\pi/4}}{\sqrt{z}}\right) + N_{1/3}\left(\frac{2e^{i\pi/4}}{\sqrt{z}}\right)\right]\Bigg\} - e^{-i\pi/4}$$
$$\times \left\{ \tan\frac{\pi}{6} \left[\mathbf{J}_{1/3}\left(\frac{2e^{-i\pi/4}}{\sqrt{z}}\right) - J_{1/3}\left(\frac{2e^{-i\pi/4}}{\sqrt{z}}\right) \right] \right.$$
$$\left.\left. - \left[\mathbf{E}_{1/3}\left(\frac{2e^{-i\pi/4}}{\sqrt{z}}\right) + N_{1/3}\left(\frac{2e^{-i\pi/4}}{\sqrt{z}}\right) \right] \right\}\right] \quad , \tag{5.7}$$

where $\mathbf{J}_\gamma(z)$, $\mathbf{E}_\gamma(z)$, $J_\gamma(z)$ and $N_\gamma(z)$ (also commonly written as $Y_\gamma(z)$) denote respectively Anger, Weber, Bessel and Neumann functions. The above result can be expressed more simply in terms of the Lommel functions $S_{\mu,\nu}(z)$:

$$S_{1/3}(a) = \frac{3\Gamma(3)}{a^3} + \frac{1}{2} - \frac{54i}{a^3} \sum_{n=1}^{\infty} \frac{1}{z^{3/2}} \left[e^{i\pi/4} S_{0,1/3}\left(\frac{2e^{i\pi/4}}{\sqrt{z}}\right) \right.$$
$$\left. - e^{-i\pi/4} S_{0,1/3}\left(\frac{2e^{-i\pi/4}}{\sqrt{z}}\right) \right] \quad . \tag{5.8}$$

Because $z \propto a^{-3}$, the series expansions for the various functions given by Nos. 8.402, 8.443 and 8.581 in Gradshteyn and Ryzhik can be used to produce a series expansion in powers of a. Thus, there are no oscillating exponentials for $S_{1/3}(a)$. To obtain the small a-series more quickly, we can therefore use the Ramanujan-Berndt result (Eq. (2.2)), which yields

$$S_{1/3}(a) = \frac{3\Gamma(3)}{a^3} + \frac{1}{2} + 2 \sum_{n=0}^{\infty} \sum_{l=1}^{5} \frac{(-1)^{l+n+1} a^{6n+l}}{(2\pi)^{2n+l/3+1}} \frac{\Gamma(2n+l/3+1)}{\Gamma(6n+l+1)}$$
$$\times \zeta(2n+l/3+1) \sin\left(\frac{l\pi}{6}\right) \quad , \tag{5.9}$$

where we have utilised the Riemann recursion relationship, No. 9.535(3) in Ref. [16]. An application of this is to replace a by at and then to multiply by $t^s e^{-t}$ where Re $s > 2$. After integrating between 0 and infinity, we get

$$\sum_{n=0}^{\infty} (an^{1/3}+1)^{-(s+1)} = \frac{3}{a^3} B(s-2,3) + \frac{1}{2} + 2 \sum_{n=0}^{\infty} \sum_{l=1}^{5} \frac{(-1)^{l+n+1} a^{6n+l}}{(2\pi)^{2n+l/3+1}}$$
$$\times \frac{\Gamma(2n+l/3+1)}{\Gamma(6n+l+1)} \Gamma(6n+l+s+1)\, \zeta(2n+l/3+1) \sin\left(\frac{l\pi}{6}\right) . \tag{5.10}$$

Of course, the procedure used to obtain Eq. (5.10) can be applied to any $S_{p/q}(a)$ that only has a zeta series.

The next case we consider is p/q equal to 1/2. Then the generalised Euler-Jacobi series becomes

$$S_{1/2}(a) = \frac{2\Gamma(2)}{a^2} + \frac{1}{2} + \frac{2}{a^2} \sum_{n=1}^{\infty} {}_2F_0\left(1, 3/2; \pm\, z e^{i\pi/2}\right)^+ \quad , \tag{5.11}$$

where z now equals $8n\pi/a^2$. Noting

$$\begin{aligned} {}_2F_0\left(1, 3/2; \pm z e^{i\pi/2}\right)^+ &= 2\int_0^\infty dt\, t\, e^{-t} \cos\left(zt^2/4\right) \\ &= \sqrt{\frac{\pi z}{8}} \int_0^\infty dy\, y^{1/2}\, e^{-\sqrt{y}}\, J_{-1/2}\left(zy/4\right) \quad , \end{aligned} \tag{5.12}$$

we can obtain two representations for $S_{1/2}(a)$. By realising that the first integral in Eq. (5.12) is related to No. 7.4.23 in Abramowitz and Stegun [17], we find that

$$\begin{aligned} S_{1/2}(a) = \frac{2\Gamma(2)}{a^2} + \frac{1}{2} + \frac{\sqrt{2\pi}}{a^2} \sum_{n=1}^{\infty} \frac{4}{z^{3/2}} \Bigg\{ \left(\frac{1}{2} - C\left(\sqrt{\frac{2}{\pi z}} \right) \right) \\ \times \cos\left(z^{-1}\right) + \left(\frac{1}{2} - S\left(\sqrt{\frac{2}{\pi z}} \right) \right) \sin\left(z^{-1}\right) \Bigg\} \quad , \end{aligned} \tag{5.13}$$

where $C(z)$ and $S(z)$ are the Fresnel integrals. To obtain the second representation, we put $\alpha = 3/2$, $\nu = -1/2$, $c = z/4$ and $p = 1$ in No. 2.12.9.4 from Prudnikov et al [25], which reproduces the second integral in Eq. (5.12). Hence, we find

$$\begin{aligned} \int_0^\infty dy\, y^{1/2} e^{-\sqrt{y}}\, J_{-1/2}(zy/4) = \frac{\sqrt{2}}{\pi} \left(\frac{4}{z}\right)^2 \Gamma(3/4)\, \Gamma(5/4) \\ \times\ {}_0F_1\left(1/2; -1/4z^2\right) - \frac{1}{\sqrt{2\pi}} \left(\frac{4}{z}\right)^{5/2} {}_1F_2\left(1; 3/4, 5/4; -1/4z^2\right) \\ + \frac{\sqrt{2}}{3\pi} \left(\frac{4}{z}\right)^3 \Gamma(5/4)\, \Gamma(7/4)\, {}_0F_1\left(3/2; -1/4z^2\right) \quad . \end{aligned} \tag{5.14}$$

The ${}_1F_2$ hypergeometric function can be expressed as a combination of Anger and Weber functions while the first and second ${}_0F_1$ hypergeometric functions can be written, respectively, as $J_{-1/2}(z^{-1})$ and $J_{1/2}(z^{-1})$. Hence, the alternative representation of $S_{1/2}(a)$ is

$$\begin{aligned} S_{1/2}(a) = \frac{2}{a^2} + \frac{1}{2} + \frac{16}{\sqrt{2}\, a^2} \sum_{n=1}^{\infty} z^{-2} \Big[\Gamma(3/4)\, \Gamma(5/4) \Big(J_{1/2}(z^{-1}) \\ + J_{-1/2}(z^{-1}) \Big) - \Big(\mathbf{J}_{1/2}(z^{-1}) + \mathbf{E}_{1/2}(z^{-1}) \Big) \Big] \quad . \end{aligned} \tag{5.15}$$

Since $z \propto 1/a^2$, we can use the series expressions for the various functions in Eq. (5.15) to obtain the small a-expansion of $S_{1/2}(a)$. By substituting these into Eq. (5.15), we get after some manipulation

$$S_{1/2}(a) = \frac{2}{a^2} + \frac{1}{2} + 8\sqrt{\frac{\pi}{2}} \sum_{n=1}^{\infty} \sum_{k=0}^{\infty} (-1)^k \left[\frac{1}{(2k+1)!} \frac{a^{4k+3}}{(8\pi n)^{2k+5/2}} + \frac{1}{(2k)!} \frac{a^{4k+1}}{(8\pi n)^{2k+3/2}} - \sum_{k=0}^{\infty} \frac{\sqrt{2}}{\Gamma(2k+3/2)} \frac{a^{4k+2}}{(8\pi n)^{2k+2}} \right] . \tag{5.16}$$

Interchanging the n and k summations and then identifying the zeta functions throughout, we finally obtain

$$S_{1/2}(a) = \frac{2}{a^2} + \frac{1}{2} + \sqrt{2} \sum_{k=0}^{\infty} (-1)^k a^{4k+1} \left[a^2 \frac{\Gamma(2k+5/2)}{\Gamma(4k+4)} \frac{\zeta(2k+5/2)}{(2\pi)^{2k+5/2}} + \frac{\Gamma(2k+3/2)}{\Gamma(4k+2)} \frac{\zeta(2k+3/2)}{(2\pi)^{2k+3/2}} - \sqrt{2}\, a \frac{\Gamma(2k+2)}{\Gamma(4k+3)} \frac{\zeta(2k+2)}{(2\pi)^{2k+2}} \right] , \tag{5.17}$$

which results from Eq. (2.2). Once again, there are no oscillating exponentials. As an aside, if we put $\alpha = at$, multiply by $t^{\gamma-1} \exp(-\alpha t - \beta/t)$ and integrate from 0 to infinity w.r.t. t, we get

$$\sum_{n=0}^{\infty} \left(\frac{\beta}{\alpha + \sqrt{n}\, a} \right)^{\gamma/2} K_{\gamma-1}\left(2(\beta(\alpha + \sqrt{n}\, a))^{1/2} \right) = \frac{2}{a^2} \left(\frac{\beta}{\alpha} \right)^{\gamma/2-1} \times K_{\gamma-2}\left(2\sqrt{\beta\alpha} \right) + \frac{1}{2} \left(\frac{\beta}{\alpha} \right)^{\gamma/2} K_{\gamma}\left(2\sqrt{\beta\alpha} \right) + \sqrt{2} \sum_{k=0}^{\infty} (-1)^k a^{4k+1} \times \left(\frac{\beta}{\alpha} \right)^{2k+(\gamma+1)/2} \left[\frac{\beta a^2}{\alpha} \frac{\Gamma(2k+5/2)}{\Gamma(4k+4)} \frac{\zeta(2k+5/2)}{(2\pi)^{2k+5/2}} K_{4k+\gamma+3}\left(2\sqrt{\beta\alpha} \right) - \frac{\Gamma(2k+3/2)}{\Gamma(4k+2)} \frac{\zeta(2k+3/2)}{(2\pi)^{2k+3/2}} K_{4k+\gamma+1}\left(2\sqrt{\beta\alpha} \right) - \sqrt{\frac{2\beta}{\alpha}}\, a \frac{\Gamma(2k+2)}{\Gamma(4k+3)} \times \frac{\zeta(2k+2)}{(2\pi)^{2k+2}} K_{4k+\gamma+2}\left(2\sqrt{\beta\alpha} \right) \right] , \tag{5.18}$$

where we have used No. 3.4771(9) from Gradshteyn and Ryzhik [16] and Re α and Re β are both greater than zero. By putting $\gamma = 1/2$, we find

$$\sum_{n=0}^{\infty} \frac{e^{-2(\beta(\alpha+\sqrt{n}\, a))^{1/2}}}{(\alpha + \sqrt{n}\, a)^{1/2}} = e^{-2\sqrt{\beta\alpha}} \left(\frac{1}{2\sqrt{\alpha}} + \frac{2\beta\sqrt{\alpha}}{a^2} + \frac{\sqrt{\beta}}{a^2} \right) + \sqrt{\frac{8}{\pi a}} \times \sum_{k=0}^{\infty} (-1)^k \left(\frac{\beta a^2}{\alpha} \right)^{2k+3/4} \left[\frac{\beta a^2}{\alpha} \frac{\Gamma(2k+5/2)}{\Gamma(4k+4)} \frac{\zeta(2k+5/2)}{(2\pi)^{2k+5/2}} \times K_{4k+7/2}\left(2\sqrt{\beta\alpha} \right) - \frac{\Gamma(2k+3/2)}{\Gamma(4k+2)} \frac{\zeta(2k+3/2)}{(2\pi)^{2k+3/2}} K_{4k+3/2}\left(2\sqrt{\beta\alpha} \right) - \sqrt{\frac{2\beta}{\alpha}}\, a \frac{\Gamma(2k+2)}{\Gamma(4k+3)} \frac{\zeta(2k+2)}{(2\pi)^{2k+2}} K_{4k+5/2}\left(2\sqrt{\beta\alpha} \right) \right] , \tag{5.19}$$

where No. 8.468 from Gradshteyn and Ryzhik [16] can be used to turn the modified Bessel functions into exponential forms involving $\exp(-2(\beta\alpha)^{1/2})$.

We complete this subsection by examining $p/q = 2/3$. Then Eq. (2.14) becomes

$$S_{2/3}(a) = \frac{3\Gamma(3/2)}{2\,a^{3/2}} + \frac{1}{2} + \frac{9}{4\sqrt{\pi}\,a^{3/2}} \sum_{n=1}^{\infty} \Big[\Gamma(5/6)\,\Gamma(7/6) \times {}_2F_0\,(5/6, 7/6; \pm\, z)^+ - z\Gamma(11/6)\,\Gamma(13/6) \times {}_3F_1\,(1, 11/6, 13/6; 2; \pm\, z)^+\Big], \tag{5.20}$$

where $z = 27n^2\pi^2/a^3$. Noting that

$$z\Gamma(11/6)\,\Gamma(13/6)\,{}_3F_1(1, 11/6, 13/6; 2; \pm\, z)^+ = \Gamma(5/6)\,\Gamma(7/6) \times {}_2F_0(5/6, 7/6; \pm\, z)^- \ ,$$

we find that Eq. (5.20) reduces to

$$S_{2/3}(a) = \frac{3\Gamma(3/2)}{2\,a^{3/2}} + \frac{1}{2} + \frac{9\Gamma(5/6)\,\Gamma(7/6)}{2\sqrt{\pi}\,a^{3/2}} \sum_{n=1}^{\infty} {}_2F_0(5/6, 7/6; -z) \ . \tag{5.21}$$

A representation for the hypergeometric function in Eq. (5.21) is not given in Prudnikov et al [22] but we can use the result on p. 257 of Erdelyi et al [26] to get

$$S_{2/3}(a) = \frac{3\Gamma(3/2)}{2\,a^{3/2}} + \frac{1}{2} + \frac{9\Gamma(5/6)\,\Gamma(7/6)}{2\sqrt{\pi}\,a^{3/2}} \sum_{n=1}^{\infty} z^{-5/6}\,\Psi(5/6, 2/3; z^{-1}) \ , \tag{5.22}$$

where Tricomi's confluent hypergeometric function $\Psi(\alpha, \beta; z)$ can be obtained from No. 9.210 of Gradshteyn and Ryzhik [16]. Hence,

$$\Psi(5/6, 2/3; z^{-1}) = \frac{\Gamma(1/3)}{\Gamma(7/6)}\,{}_1F_1(5/6; 2/3; z^{-1}) + \frac{\Gamma(-1/3)}{\Gamma(5/6)}\,z^{-1/3} \times {}_1F_1(7/6; 4/3; z^{-1}) \ . \tag{5.23}$$

A small a-series, therefore, emerges once Eq. (5.23) is introduced into Eq. (5.22). Hence, we arrive at

$$S_{2/3}(a) = \frac{3\Gamma(3/2)}{2a^{3/2}} + \frac{1}{2} + \sum_{k=0}^{\infty} \left[\frac{(a/3)^{3k+1}}{\pi^{2k+7/6}} \frac{\Gamma(k+5/6)}{\Gamma(k+1)} \frac{\zeta(2k+5/3)}{\Gamma(k+2/3)} - \frac{(a/3)^{3k+2}}{\pi^{2k+11/6}} \frac{\Gamma(k+7/6)}{\Gamma(k+1)} \frac{\zeta(2k+7/3)}{\Gamma(k+4/3)} \right] \ , \tag{5.24}$$

which is just the Ramanujan-Berndt result. Finally, putting $a = at^p$ and then multiplying $t^{\gamma-1}\exp(-\mu t^p)$, we get after integrating w.r.t. t from 0 to infinity

$$\Gamma(\gamma)\sum_{n=0}^{\infty}(an^{2/3}+\mu)^{-\gamma} = \frac{3\Gamma(3/2)\,\Gamma(\gamma-3/2)}{2a^{3/2}\mu^{\gamma-3/2}} + \frac{\Gamma(\gamma)}{2\mu^{\gamma}}$$
$$+\sum_{k=0}^{\infty}\left[\frac{(a/3)^{3k+1}}{\pi^{2k+7/6}}\,\frac{\Gamma(k+5/6)}{\Gamma(k+1)}\,\frac{\Gamma(3k+\gamma+1)}{\Gamma(k+2/3)}\,\frac{\zeta(2k+5/3)}{\mu^{3k+\gamma+1}}\right.$$
$$\left.-\frac{(a/3)^{3k+2}}{\pi^{2k+11/6}}\,\frac{\Gamma(k+7/6)}{\Gamma(k+1)}\,\frac{\Gamma(3k+\gamma+2)}{\Gamma(k+4/3)}\,\frac{\zeta(2k+7/3)}{\mu^{3k+\gamma+2}}\right] , \tag{5.25}$$

where we have replaced γ/p by γ and have used No. 3.478(1) from Gradshteyn and Ryzhik [16]. This result is only valid for $\gamma > 3/2$ and Re $\mu > 0$.

B. $p/q = 1$

As we mentioned earlier, before we evaluate $S_1(a)$ via Eq. (2.14) or (2.16), we shall consider evaluating $S_1^{r,m}(a)$ since $p/q = 1$ is the simplest case to evaluate for the generalised Euler-Jacobi series and hence, should be the simplest case for $S_1^{r,m}(a)$. Furthermore, Berndt has shown in Ref. [1] that $S_1^{1,1}(a)$ is given by

$$S_1^{1,1}(a) = \sum_{n=1}^{\infty} e^{-an}\log n \sim \frac{-\gamma-\log a}{a} + \frac{1}{2}\log 2\pi \ , \tag{5.26}$$

where γ is once again Euler's constant.

For $p/q \le 1$, the evaluation of $S_{p/q}^{r}(a)$ is easily accomplished when $r = 1 + mp/q$ with $m \in \mathbf{Z}^+$ since

$$S_{p/q}^{1+mp/q}(a) = \sum_{n=1}^{\infty} n^{mp/q} e^{-an^{p/q}} = (-1)^m \frac{d^m}{da^m}\, S_{p/q}(a) \ . \tag{5.27}$$

For example, by using Eq. (5.24), one finds that

$$\sum_{n=1}^{\infty} n^{2m/3} e^{-an^{2/3}} = \frac{3\Gamma(m+3/2)}{a^{m+3/2}} + \sqrt{3}\,(-1)^m \sum_{k\ge(m-1)/3}^{\infty} \frac{a^{3k-m+1}}{(2\pi)^{2k+5/3}}$$
$$\times\ \frac{\Gamma(2k+5/3)}{\Gamma(3k-m+2)}\,\zeta(2k+5/3) + \sqrt{3}\,(-1)^{m+1} \sum_{k\ge(m-2)/3}^{\infty} \frac{a^{3k-m+2}}{(2\pi)^{2k+7/3}}$$
$$\times\ \frac{\Gamma(2k+7/3)}{\Gamma(3k-m+3)}\,\zeta(2k+7/3) \ . \tag{5.28}$$

The situation, however, is different when $r \neq 1 + mp/q$. Then we have to turn either to Eq. (2.13) or to Eq. (2.15), which yield for $p/q = 1$ and $s = 0$

$$S_1^{r}(a) = \frac{\Gamma(r)}{a^r} + \frac{1}{2}\,\delta_{r,1} + \frac{1}{a}\sum_{n=1}^{\infty} a^{1-r}\left[\Gamma(r+1)\,{}_2F_1\!\left(1, r+1; 2; \pm\left(\frac{2n\pi e^{i\pi/2}}{a}\right)\right)^{+}\right.$$

$$- (r-1)\,\Gamma(r)\,{}_2F_1\Big(1, r; 2; \pm\Big(\frac{2n\pi e^{i\pi/2}}{a}\Big)\Big)^+\Bigg] = \frac{\Gamma(r)}{a^r} + \frac{1}{2}\,\delta_{r,1}$$

$$+ \frac{1}{a}\sum_{n=1}^{\infty} a^{1-r}\left[G_{2,2}^{1,2}\left(\pm\, z e^{i\pi/2}\,\middle|\,\begin{matrix}0,-r\\0,-1\end{matrix}\right)^+ - (r-1)\right.$$

$$\left.\times\, G_{2,2}^{1,2}\left(\pm\, z e^{i\pi/2}\,\middle|\,\begin{matrix}0,1-r\\0,-1\end{matrix}\right)^+\right] \,. \tag{5.29}$$

In the second representation, $|z| = 2n\pi/a < 1$.

We can simplify Eq. (5.29) by noting that for $|z| < 1$

$$G_{2,2}^{1,2}\left(\pm\, z e^{i\pi/2}\,\middle|\,\begin{matrix}0,-r\\0,-1\end{matrix}\right)^+ = \frac{1}{iz}\sum_{k=0}^{\infty}\frac{\Gamma(k+r)}{\Gamma(k+1)}\, z^k\left(e^{i\pi k/2} - e^{-i\pi k/2}\right)$$

$$= e^{-i\pi/2} z^{-1}\left((1-iz)^{-r} - (1+iz)^{-r}\right)\Gamma(r) = 2\Gamma(r)\, z^{-1}$$

$$\times\,\left(1+z^2\right)^{-r/2}\sin(r\arctan z) \,, \tag{5.30}$$

and

$$G_{2,2}^{1,2}\left(\pm\, z e^{i\pi/2}\,\middle|\,\begin{matrix}0,1-r\\0,-1\end{matrix}\right)^+ = \left(\frac{\Gamma(r-1)}{iz}\right)\Big((1-iz)^{1-r}$$

$$- (1+iz)^{1-r}\Big) = 2\Gamma(r-1)\, z^{-1}\left(1+z^2\right)^{(1-r)/2}$$

$$\times\,\sin\Big((1-r)\arctan z\Big) \,. \tag{5.31}$$

For $r = 1$, Eq. (5.29) becomes the generalised Euler-Jacobi series with $p/q = 1$ and as a consequence, we should obtain Eq. (2.5). Putting $r = 1$ means that the contribution from the second Meijer G-function in Eq. (5.29) vanishes while Eq. (5.30) yields for $a > 0$

$$S_1(a) = \frac{1}{a} + \frac{1}{2} + \sum_{n=1}^{\infty}\frac{(a/2\pi^2)}{a^2/4\pi^2 + n^2} = \frac{1}{1-e^{-a}} \,, \tag{5.32}$$

where we have utilised Mittag-Leffler's partial fraction decomposition formula [11] for $\coth(a/2)$ to obtain Eq. (5.32).

The introduction of Eqs. (5.30) and (5.31) into Eq. (5.29) gives

$$S_1^r(a) = \frac{\Gamma(r)}{a^r} + \frac{1}{2}\,\delta_{r,1} + 2\sum_{n=1}^{\infty}\frac{a^{-r}\Gamma(r)}{z(1+z^2)^{r/2}}\Big(2\sin(r\arctan z)$$

$$- \; z\cos(r\arctan z)\Big) \,. \tag{5.33}$$

Taking the first and second derivatives of the above equation w.r.t. r yields

$$S_1^{r,1}(a) = a^{-r}\left(\Psi(r)\,\Gamma(r) - \Gamma(r)\log a\right) + 2\,\Gamma(r) \times \sum_{n=1}^{\infty} \frac{(\Psi(r) - f(a,z))}{(a^2+a^2z^2)^{r/2}} \Big(2\sin(r\arctan z) - z\cos(r\arctan z)\Big) + 2\sum_{n=1}^{\infty} \frac{\Gamma(r)\arctan z}{(a^2+a^2z^2)^{r/2}} \Big(2\cos(r\arctan z) + z\sin(r\arctan z)\Big) \;, \tag{5.34}$$

and

$$S_1^{r,2}(a) = a^{-r}\,\Gamma(r)\Big(\Psi'(r) + \Psi(r)(\Psi(r) - 2\log a) + \log^2 a\Big) + 2\sum_{n=1}^{\infty}\left(\frac{\Gamma(r)(\Psi(r)^2 + \Psi'(r)) - f(a,z))}{(a^2+a^2z^2)^{r/2}}\right)\Big(2\sin(r\arctan z) - z\cos(r\arctan z)\Big) + 2\Gamma(r)\sum_{n=1}^{\infty}\left(\frac{(\Psi(r) - f(a,z))\arctan z}{(a^2+a^2z^2)^{r/2}}\right) \times \Big(2\cos(r\arctan z) + z\sin(r\arctan z)\Big) + 2\sum_{n=1}^{\infty}\frac{\arctan z}{(a^2+a^2z^2)^{r/2}} \times \Big[\Psi(r)\Gamma(r) - f(a,z) + \arctan z\,\Big(z\cos(r\arctan z) - 2\sin(r\arctan z)\Big)\Big] \;, \tag{5.35}$$

where $f(a,z) = (1/2)\log(a^2 + a^2z^2)$. Now putting $r = 1$ in Eq. (5.34) yields

$$S_1^{1,1}(a) = \frac{-\gamma - \log a}{a} + 2\sum_{n=1}^{\infty}\Big[\Big(\gamma - (1/2)\log(a^2 + 4n^2\pi^2)\Big) \times \frac{2n\pi}{a^2+4n^2\pi^2} + \arctan(2n\pi/a)\left(\frac{1}{a} + \frac{a}{a^2+4n^2\pi^2}\right)\Big] \;, \tag{5.36}$$

which is another representation of Berndt's result Eq. (5.26).

C. $1 < p/q < 2$

We begin this subsection by considering the generalised Euler-Jacobi series for $p/q = 4/3$. Then according to Eq. (2.16), the series can be written as

$$S_{4/3}(a) = \frac{3\,\Gamma(3/4)}{4\,a^{3/4}} + \frac{1}{2} + \frac{3^{5/4}\sqrt{2\pi}}{8\,a^{3/4}}\sum_{n=1}^{\infty}\sum_{l=0}^{3}(-1)^l z^{l/2} \times G_{3,4}^{1,3}\left(\pm z\,\middle|\,\begin{matrix}0, 1/3-(2l+1)/4, 2/3-(2l+1)/4\\ 0, -l/2, 1/4-l/2, 1/2-l/2\end{matrix}\right)^{+} \;, \tag{5.37}$$

where $z = (3/a)^3(n\pi/2)^4$. To evaluate Eq. (5.37), we require

$$G_{3,4}^{1,3}\left(\pm z\,\middle|\,\begin{matrix}0, 1/12, 5/12\\ 0, 0, 1/4, 1/2\end{matrix}\right)^{+} = \frac{\Gamma(7/12)}{\Gamma(1/2)}\frac{\Gamma(11/12)}{\Gamma(3/4)} \times {}_2F_2\!\left(\frac{7}{12}, \frac{11}{12}; \frac{1}{2}, \frac{3}{4}; \pm z\right)^{+} \;, \tag{5.38}$$

$$G^{1,3}_{3,4}\left(\pm z \left| \begin{matrix} 0, -5/12, -1/12 \\ 0, -1/2, -1/4, 0 \end{matrix} \right.\right)^{+} = \frac{\Gamma(13/12)}{\Gamma(5/4)} \frac{\Gamma(17/12)}{\Gamma(3/2)} \times {}_2F_2\left(\frac{13}{12}, \frac{17}{12}; \frac{5}{4}, \frac{3}{2}; \pm z\right)^{+}, \tag{5.39}$$

$$G^{1,3}_{3,4}\left(\pm z \left| \begin{matrix} 0, -11/12, -7/12 \\ 0, -1, -3/4, -1/2 \end{matrix} \right.\right)^{+} = z^{-1} \frac{\Gamma(7/12)}{\Gamma(1/2)} \frac{\Gamma(11/12)}{\Gamma(3/4)} \times {}_2F_2\left(\frac{7}{12}, \frac{11}{12}; \frac{1}{2}, \frac{3}{4}; \pm z\right)^{-}, \tag{5.40}$$

$$G^{1,3}_{3,4}\left(\pm z \left| \begin{matrix} 0, -17/12, -13/12 \\ 0, -3/2, -5/4, -1 \end{matrix} \right.\right)^{+} = z^{-1} \frac{\Gamma(13/12)}{\Gamma(5/4)} \frac{\Gamma(17/12)}{\Gamma(3/2)} \times {}_2F_2\left(\frac{13}{12}, \frac{17}{12}; \frac{5}{4}, \frac{3}{2}; \pm z\right)^{-}. \tag{5.41}$$

Introducing Eqs. (5.38) to (5.41) into Eq. (5.37) gives

$$S_{4/3} = \frac{3\Gamma(3/4)}{4\, a^{3/4}} + \frac{1}{2} + \frac{3^{5/4}\sqrt{2\pi}}{4\, a^{3/4}} \sum_{n=1}^{\infty} \left[\frac{\Gamma(7/12)}{\Gamma(1/2)} \frac{\Gamma(11/12)}{\Gamma(3/4)} \times {}_2F_2\left(\frac{7}{12}, \frac{11}{12}; \frac{1}{2}, \frac{3}{4}; z\right) - z^{1/2} \frac{\Gamma(13/12)}{\Gamma(5/4)} \frac{\Gamma(17/12)}{\Gamma(3/2)} \times {}_2F_2\left(\frac{13}{12}, \frac{17}{12}; \frac{5}{4}, \frac{3}{2}; z\right) \right]. \tag{5.42}$$

Although many forms for ${}_2F_2$ hypergeometric functions appear in Prudnikov et al [22], those given in Eq. (5.42) do not. Hence, we need to apply the asymptotics presented in Sec. 3 to obtain a small a-expansion for $S_{4/3}(a)$.

From Eqs. (3.16) and (3.17), we obtain

$${}_2F_2\left(\frac{7}{12}, \frac{11}{12}; \frac{1}{2}, \frac{3}{4}; z\right) \sim \frac{\Gamma(1/2)}{\Gamma(7/12)} \frac{\Gamma(3/4)}{\Gamma(11/12)} \times \left[K_{2,2}(z) + L_{2,2}(ze^{\mp i\pi}) \right], \tag{5.43}$$

and

$${}_2F_2\left(\frac{13}{12}, \frac{17}{12}; \frac{5}{4}, \frac{3}{2}; z\right) \sim \frac{\Gamma(5/4)}{\Gamma(13/12)} \frac{\Gamma(3/2)}{\Gamma(17/12)} \times \left[K_{2,2}(z) + L_{2,2}(ze^{\mp i\pi}) \right], \tag{5.44}$$

where the '−' applies when $\delta - \pi/2 \le \arg z \le 3\pi/2 - \delta$ ($\delta > 0$) and the '+' applies when $\delta - 3\pi/2 \le \arg z \le \pi/2 - \delta$. As mentioned earlier, the apparently conflicting result for $|\arg z| < \pi/2$ arises as a consequence of the Stokes phenomenon. We shall, therefore, evaluate the asymptotics for the ${}_2F_2$ hypergeometric functions using both expressions for the L-asymptotics.

Now using Eqs. (3.2) and (3.3), we get for the K- and L-asymptotics

$$K_{2,2}(z) = z^{1/4}\, e^{z} \sum_{r=0}^{\infty} N_r \begin{pmatrix} 7/12 & 11/12 \\ 1/2 & 3/4 \end{pmatrix} z^{-r} \;, \tag{5.45}$$

$$K_{2,2}(z) = z^{-1/4}\, e^{z} \sum_{r=0}^{\infty} N_r \begin{pmatrix} 13/12 & 17/12 \\ 5/4 & 3/2 \end{pmatrix} z^{-r} \;, \tag{5.46}$$

$$\begin{aligned} L_{2,2}\left(ze^{\pm i\pi}\right) &= \left(ze^{\pm i\pi}\right)^{-7/12} \frac{\Gamma(7/12)}{\Gamma(-1/12)} \frac{\Gamma(1/3)}{\Gamma(1/6)} \\ &\times {}_3F_1\left(\begin{matrix} 7/12, 5/6, 13/12 \\ 2/3 \end{matrix} \middle| \frac{e^{\mp i\pi}}{z} \right) + \left(ze^{\pm i\pi}\right)^{-11/12} \frac{\Gamma(11/12)}{\Gamma(-5/12)} \\ &\times \frac{\Gamma(-1/3)}{\Gamma(-1/6)}\, {}_3F_1\left(\begin{matrix} 11/12, 7/6, 17/12 \\ 4/3 \end{matrix} \middle| \frac{e^{\mp i\pi}}{z} \right) \;, \end{aligned} \tag{5.47}$$

and

$$\begin{aligned} L_{2,2}\left(ze^{\pm i\pi}\right) &= \left(ze^{\pm i\pi}\right)^{-13/12} \frac{\Gamma(13/12)}{\Gamma(1/6)} \frac{\Gamma(1/3)}{\Gamma(5/12)} \\ &\times {}_3F_1\left(\begin{matrix} 7/12, 5/6, 13/12 \\ 2/3 \end{matrix} \middle| \frac{e^{\mp i\pi}}{z} \right) + \left(ze^{\pm i\pi}\right)^{-17/12} \frac{\Gamma(17/12)}{\Gamma(-1/6)} \\ &\times \frac{\Gamma(-1/3)}{\Gamma(1/12)}\, {}_3F_1\left(\begin{matrix} 11/12, 7/6, 17/12 \\ 4/3 \end{matrix} \middle| \frac{e^{\mp i\pi}}{z} \right) \;, \end{aligned} \tag{5.48}$$

where Eqs. (5.47) and (5.48) pertain to the hypergeometric functions given in Eqs. (5.43) and (5.44), respectively. As mentioned in Sec. 3, the N_r in Eqs. (5.45) and (5.46) are evaluated via the recursion relations given by Eqs. (3.5)-(3.12). Furthermore, since $S_{4/3}(a)$ is convergent, the recursion relation for the N_r, which are dependent upon the indices or parameters of the ${}_2F_2$ hypergeometric functions, must be identical to cancel the exponentially diverging series given in Eqs. (5.45) and (5.46).

As can be seen from Eqs. (3.5) and (3.7), the N_r are expressed in terms of the c_r, which can only be determined by evaluating the $T(t)$ and $U(t)$ in Eqs. (3.9) and (3.10), respectively. In turn, the $T(t)$ and $U(t)$ can only be determined once the ω_j and λ_j given by Eqs. (3.13) and (3.14) are evaluated. The ω_j and λ_j are dependent on the indices (parameters) of a hypergeometric function and a cursory look at the ${}_2F_2$ hypergeometric functions in Eq. (5.42) would seem to suggest that the ω_j and λ_j, and ultimately the corresponding N_r for each hypergeometric function are different. However, whilst each of the ω_j and λ_j for the hypergeometric functions are different, the $T(t)$ and $U(t)$ are not since these are given respectively by products involving all the ω_j and λ_j. Thus, for both the hypergeometric functions in Eqs. (5.42) we find that $\{\omega_j\} = \{0, \pm 1/4\}$ and $\{\lambda_j\} = \{5/6, 7/6\}$ since $\rho_0 = 1$. This means that

$$T(t) = (t + 1/4)(t - 1/4)t \ , \tag{5.49}$$

and

$$U(t) = (t + 5/6)(t + 7/6) \ . \tag{5.50}$$

Introducing these results into Eqs. (3.11) and (3.12), we find that the K-asymptotics for the first hypergeometric function in Eq. (5.42) can be written as

$$\begin{aligned} {}_2F_2\left(\frac{7}{12}, \frac{11}{12}; \frac{1}{2}, \frac{3}{4}; z\right)_K &\sim \frac{\Gamma(1/2)}{\Gamma(7/12)} \frac{\Gamma(3/4)}{\Gamma(11/12)} \, z^{1/4} \, e^z \\ &\times \left(1 - \frac{5}{144z} + O\left(z^{-2}\right)\right) \ , \end{aligned} \tag{5.51}$$

while those for the second become

$$\begin{aligned} {}_2F_2\left(\frac{13}{12}, \frac{17}{12}; \frac{5}{4}, \frac{3}{2}; z\right)_K &\sim \frac{\Gamma(13/12)}{\Gamma(5/4)} \frac{\Gamma(17/12)}{\Gamma(3/2)} \, z^{-1/4} \, e^z \\ &\times \left(1 - \frac{5}{144z} + O\left(z^{-2}\right)\right) \ , \end{aligned} \tag{5.52}$$

where the subscript K denotes that only the K-asymptotics have been considered. Upon introducing this into the generalised Euler-Jacobi series for $p/q = 4/3$, i.e. Eq. (5.42), we find that the K-asymptotics for both hypergeometric functions cancel each other completely.

The complete cancellation of the K-asymptotics is due to the equality of the $T(t)$ and $U(t)$ for both hypergeometric functions. The fact that hypergeometric functions with different indices (parameters) can produce identical K-asymptotics except for a power of z (in the above case it was $z^{1/2}$), is attributed to the extra degree of freedom introduced by having to calculate ω_0 for the $T(t)$. Whereas all the ω_j are directly dependent on each ρ_j or lower parameter of a hypergeometric function (see Eq. (3.13)), ω_0 depends only on β, γ and ρ_0, the last of these quantities equalling unity. Hence, the appearance of ω_0 in $T(t)$ enables the ω_j to be permuted cyclically. For instance $\{\omega_0 = 1/4, \omega_1 = -1/4, \omega_2 = 0\}$ were the ω_j values obtained for ${}_2F_2(7/12, 11/12; 1/2, 3/4; z)$ while for ${}_2F_2(13/12, 17/12; 5/4, 3/2; z)$ the ω_j values were $\{\omega_0 = -1/4, \omega_1 = 0, \omega_2 = 1/4\}$. In addition, the hypergeometric function, resulting from the one remaining cyclic permutation of the ω_j, $\{\omega_0 = 0, \omega_1 = 1/4, \omega_2 = -1/4\}$ should produce the same K-asymptotics except for a power of z. The hypergeometric function producing this last combination of ω_js is ${}_2F_2(5/6,7/6;3/4,5/4;z)$ whilst the power of z required to make the K-asymptotic part equal those of ${}_2F_2(7/12,11/12;1/2,3/4;z)$ is $z^{1/4}$. Hence, the asymptotic expansion of ${}_2F_2(5/6,7/6;3/4,5/4;z)$ can be immediately determined to be

$$
\begin{aligned}
{}_2F_2\left(\frac{5}{6},\frac{7}{6};\frac{3}{4},\frac{5}{4};z\right) &\sim \frac{\Gamma(3/4)}{\Gamma(5/6)}\frac{\Gamma(5/4)}{\Gamma(7/6)}\left\{e^z\left(1-\frac{5}{144z}\right.\right.\\
&\left.+O\left(z^{-2}\right)\right)+\left(ze^{\pm i\pi}\right)^{-5/6}\frac{\Gamma(5/6)}{\Gamma(-1/12)}\frac{\Gamma(1/3)}{\Gamma(5/12)}\\
\times\,{}_3F_1\left(\frac{5}{6},\frac{7}{12},\frac{13}{12};\frac{2}{3};\frac{e^{\mp i\pi}}{z}\right) &+\left(ze^{\pm i\pi}\right)^{-7/6}\frac{\Gamma(7/6)}{\Gamma(-5/12)}\frac{\Gamma(-1/3)}{\Gamma(1/12)}\\
&\left.\times\,{}_3F_1\left(\frac{7}{6},\frac{11}{12},\frac{17}{12};\frac{4}{3};\frac{e^{\mp i\pi}}{z}\right)\right\}, \qquad (5.53)
\end{aligned}
$$

where we have used Eq. (3.3).

So far, we have shown that there are no exponential terms appearing in the asymptotic expansion of the generalised Euler-Jacobi series for $p/q = 4/3$. We still require the L-asymptotics, which we already know from Sec. 3 must produce the zeta series obtained by Ramanujan and Berndt. From Eqs. (5.43), (5.44), (5.47) and (5.48), we find that after some algebra Eq. (5.42) becomes

$$
\begin{aligned}
S_{4/3}(a) &= \frac{3\Gamma(3/4)}{4a^{3/4}}+\frac{1}{2}+\frac{3^{5/4}\sqrt{2\pi}}{4\,a^{3/4}}\sum_{n=1}^{\infty}\left[\frac{\Gamma(7/12)\,\Gamma(13/12)}{z^{7/12}\,\pi}\right.\\
\times\,\frac{\Gamma(1/3)}{\Gamma(1/6)}\,{}_3F_1\left(\frac{7}{12},\frac{5}{6},\frac{13}{12};\frac{2}{3};\frac{e^{\mp i\pi}}{z}\right) &+\frac{\Gamma(11/12)\,\Gamma(17/12)}{z^{11/12}\,\pi}\frac{\Gamma(-1/3)}{\Gamma(-1/6)}\\
&\left.\times\,{}_3F_1\left(\frac{11}{12},\frac{7}{6},\frac{17}{12};\frac{4}{3};\frac{e^{\mp i\pi}}{z}\right)\right]. \qquad (5.54)
\end{aligned}
$$

Here we can drop the $\mp$, since either way the result for $S_{4/3}(a)$ will be the same. Thus, the Stokes phenomenon has no effect on Eq. (5.54). This has arisen because in doing the algebra to get to Eq. (5.54), we were consistent in using either all the upper signs for the ${}_2F_2$ hypergeometric functions in Eq. (5.42) or the lower signs for the ${}_2F_2$ hypergeometric functions. Had we used the 'upper sign' version for the first ${}_2F_2$ in Eq. (5.42) and the 'lower sign' version for the second ${}_2F_2$ in the same equation, which is allowed since both results are valid for $|\arg z| < \pi/2$, then our answer for $S_{4/3}(a)$ would have been complex. A complex result would also have arisen had we used the lower sign and upper sign versions for the first and second ${}_2F_2$ hypergeometric functions, respectively, in Eq. (5.42). Hence, we can appeal to the reality of $S_{4/3}(a)$ to remove the problem associated with the Stokes phenomenon.

If we utilise the series expansion for the ${}_3F_1$ hypergeometric functions in Eq. (5.54) and the Gauss multiplication formula for the gamma function, then Eq. (5.54) becomes

$$S_{4/3}(a) = \frac{3\Gamma(3/4)}{4a^{3/4}} + \frac{1}{2} + \frac{3^{5/4}}{2a^{3/4}} \sum_{n=1}^{\infty} \left[\frac{z^{-7/12}}{4^{5/6}} \sum_{k=0}^{\infty} (-1)^k \frac{\Gamma(4k+4/3)}{\Gamma(3k+1)} \right.$$
$$\left. \times \left(\frac{27}{256z}\right)^k + \frac{z^{-11/12}}{4^{13/6}} \sum_{k=0}^{\infty} (-1)^k \frac{\Gamma(4k+8/3)}{\Gamma(3k+2)} \left(\frac{27}{256\,z}\right)^k \right] . \tag{5.55}$$

Interchanging the sums, we finally get

$$S_{4/3}(a) = \frac{3\Gamma(3/4)}{4a^{3/4}} + \frac{1}{2} + \frac{4}{\sqrt{3}} \sum_{k=0}^{\infty} \left[(-1)^k \frac{a^{3k+1}}{(2\pi)^{4k+4/3}} \Gamma(4k+4/3) \right.$$
$$\left. \times \zeta(4k+7/3) + (-1)^k \frac{a^{3k+2}}{(2\pi)^{4k+11/3}} \Gamma(4k+8/3)\, \zeta(4k+11/3) \right] , \tag{5.56}$$

which is the result obtained by using Eq. (2.2).

We now evaluate the generalised Euler-Jacobi series for $p/q = 3/2$ to see if the cancellation of the K-asymptotics occurs leaving us only to evaluate the L-asymptotics, which is equivalent to using the Ramanujan-Berndt result to determine the small a-series. Thus, according to Eq. (2.16), the generalised Euler-Jacobi series becomes

$$S_{3/2}(a) = \frac{2\Gamma(2/3)}{3a^{2/3}} + \frac{1}{2} + \frac{2^{5/3}}{3\,a^{2/3}} \sqrt{\frac{\pi}{3}} \sum_{n=1}^{\infty} \sum_{l=0}^{2} (-1)^l \, z^{2l/3}$$
$$\times\ G_{2,3}^{1,2}\left(\pm\, z e^{i\pi/2} \,\middle|\, \begin{matrix} 0, 1/2-(2l+1)/3 \\ 0, -2l/3, 1/3-2l/3 \end{matrix} \right)^{+} , \tag{5.57}$$

where $z = (2/a)^2 (2n\pi/3)^3$. In order to evaluate this, we need

$$G_{2,3}^{1,2}\left(\pm\, iz \,\middle|\, \begin{matrix} 0, 1/6 \\ 0, 0, 1/3 \end{matrix} \right)^{+} = \frac{\Gamma(5/6)}{\Gamma(2/3)}\, {}_1F_1\left(\frac{5}{6}; \frac{2}{3}; \pm\, iz\right)^{+} , \tag{5.58}$$

$$G_{2,3}^{1,2}\left(\pm\, iz \,\middle|\, \begin{matrix} 0, -1/2 \\ 0, -2/3, -1/3 \end{matrix} \right)^{+} = \frac{\Gamma(3/2)}{\Gamma(4/3)\Gamma(5/3)}\, {}_2F_2\left(1, \frac{3}{2}; \frac{4}{3}, \frac{5}{3}; \pm\, iz\right)^{+} , \tag{5.59}$$

$$G_{2,3}^{1,2}\left(\pm\, iz \,\middle|\, \begin{matrix} 0, -7/6 \\ 0, -4/3, -1 \end{matrix} \right)^{+} = z^{-1} \frac{\Gamma(7/6)}{\Gamma(4/3)}\, {}_1F_1\left(\frac{7}{6}; \frac{4}{3}; \pm\, iz\right)^{-} . \tag{5.60}$$

Although Prudnikov et al [22] exhibit many representations for ${}_1F_1$ and ${}_2F_2$ hypergeometric functions, those with the specific parameters appearing in Eqs. (5.58)-(5.60) are not among these. We shall therefore identify them here.

Eq. (5.58) can be written as

$$G_{2,3}^{1,2}\left(\pm\, iz \,\middle|\, \begin{matrix} 0, 1/6 \\ 0, 0, 1/3 \end{matrix} \right)^{+} = \int_0^{\infty} dt\, e^{-t}\, t^{-1/6} \sum_{k=0}^{\infty} \frac{(zt)^k}{k!} \frac{\left(e^{i\pi k/2} + e^{-i\pi k/2}\right)}{\Gamma(k-1/3+1)} , \tag{5.61}$$

where we have introduced the integral representation for $\Gamma(k+5/6)$ and have also invoked absolute convergence to interchange the sum and integration. Noting that both series in Eq. (5.61) are Bessel functions, we get

$$G^{1,2}_{2,3}\left(\pm iz \left| \begin{matrix} 0, 1/6 \\ 0, 0, 1/3 \end{matrix} \right.\right)^{+} = \sqrt{2} \int_0^\infty dt\, z^{1/6}\, e^{-t}$$
$$\times \left[\mathrm{ber}_{-1/3}\left(2\sqrt{zt}\right) - \mathrm{bei}_{-1/3}\left(2\sqrt{zt}\right)\right] , \tag{5.62}$$

where the Kelvin functions have been introduced because

$$J_{-1/3}\left(xe^{-i\pi/4}\right) = e^{\pi i/3}\left(\mathrm{ber}_{-1/3}\, x + i\,\mathrm{bei}_{-1/3}\, x\right) , \tag{5.63}$$

and

$$J_{-1/3}\left(xe^{i\pi/4}\right) = e^{-\pi i/3}\left(\mathrm{ber}_{-1/3}\, x - i\,\mathrm{bei}_{-1/3}\, x\right) . \tag{5.64}$$

Making the substitution $y = zt$ and utilising Nos. 6.872(1) and 6.872(2) from Gradshteyn and Ryzhik [16], we find that

$$G^{1,2}_{2,3}\left(\pm iz \left| \begin{matrix} 0, 1/6 \\ 0, 0, 1/3 \end{matrix} \right.\right)^{+} = \sqrt{\pi}\, z^{2/3}\left[J_{-2/3}(z/2)\cos(z/2)\right.$$
$$\left. - J_{1/3}(z/2)\sin(z/2)\right] . \tag{5.65}$$

Eq. (5.59) can be written as

$$G^{1,2}_{2,3}\left(\pm iz \left| \begin{matrix} 0, -1/2 \\ 0, -2/3, -1/3 \end{matrix} \right.\right)^{+} = \int_0^\infty dt\, t^{1/2}\, e^{-t}$$
$$\times \sum_{k=0}^{\infty} \frac{\left(\left(zte^{i\pi/4}\right)^{2k} + \left(zte^{-i\pi/4}\right)^{2k}\right)}{\Gamma(k+3/2-1/6)\,\Gamma(k+3/2+1/6)} , \tag{5.66}$$

where we have followed a similar procedure to that employed to evaluate the previous Meijer G-function. Noting that both the series in Eq. (5.66) can be related to Anger-Weber functions, we can write Eq. (5.66) as

$$G^{1,2}_{2,3}\left(\pm iz \left| \begin{matrix} 0, -1/2 \\ 0, -2/3, -1/3 \end{matrix} \right.\right)^{+} = \frac{1}{2\sqrt{z}} \int_0^\infty dt\, e^{-t}$$
$$\times \left[e^{i\pi/4}\left(\mathbf{J}_{1/3}\left(2\sqrt{zt}\, e^{-i\pi/4}\right) - \sqrt{3}\,\mathbf{E}_{1/3}\left(2\sqrt{zt}\, e^{-i\pi/4}\right)\right)\right.$$
$$\left. + e^{-i\pi/4}\left(\mathbf{J}_{1/3}\left(2\sqrt{zt}\, e^{i\pi/4}\right) - \sqrt{3}\,\mathbf{E}_{1/3}\left(2\sqrt{zt}\, e^{i\pi/4}\right)\right)\right] . \tag{5.67}$$

Although the above integrals are given as No. 2.8.3.2 in Prudnikov et al [22], they only yield our original ${}_2F_2$ hypergeometric functions as in Eq. (5.59).

Eq. (5.60) can be written as

$$G_{2,3}^{1,2}\left(\pm ze^{i\pi/2}\,\middle|\,\begin{matrix}0,-7/6\\0,-4/3,-1\end{matrix}\right)^{+} = (iz)^{-1}\int_0^\infty dt\; z^{-1/6}\, e^{-t}$$
$$\times\; \left[e^{-i\pi/12} J_{1/3}\left(2\sqrt{zt}\, e^{i\pi/4}\right) - e^{i\pi/12} J_{1/3}\left(2\sqrt{zt}\, e^{-i\pi/4}\right)\right] \; . \tag{5.68}$$

Identifying the Bessel functions as Kelvin functions and again using Nos. 6.872(1) and 6.872(2) from Ref. [16], we find that Eq. (5.68) becomes

$$G_{2,3}^{1,2}\left(\pm ze^{i\pi/2}\,\middle|\,\begin{matrix}0,-7/6\\0,-4/3,-1\end{matrix}\right)^{+} = -\sqrt{\pi}\, z^{-2/3}\left[J_{-1/3}(z/2)\sin(z/2)\right.$$
$$\left. -\; J_{2/3}(z/2)\cos(z/2)\right] \; . \tag{5.69}$$

If we introduce Eqs. (5.59), (5.65) and (5.69) into our result for $p/q = 3/2$, then we get

$$S_{3/2}(a) = \frac{2\Gamma(2/3)}{3\,a^{2/3}} + \frac{1}{2} + \frac{2^{5/3}}{3\sqrt{3}\,a^{2/3}}\sum_{n=1}^{\infty} z^{2/3}\left[\pi\left(\cos(z/2)\right.\right.$$
$$\times\; \left(J_{-2/3}(z/2) \;-\; J_{2/3}(z/2)\right) - \sin(z/2)\left.\left(J_{1/3}(z/2) + J_{-1/3}(z/2)\right)\right)$$
$$\left. +\; \frac{9\sqrt{3}}{8}\; {}_2F_2\left(1, 3/2; 4/3, 5/3; \pm iz\right)^{+}\right] \; , \tag{5.70}$$

where the ${}_2F_2$ hypergeometric functions can be replaced by the integral in Eq. (5.67) if required. Although this result is interesting because of its connection with known functions, the small a-series is best determined by evaluating the L- and K-asymptotics of each of the hypergeometric functions in Eqs. (5.58)-(5.60).

We evaluate the L-asymptotics for the ${}_2F_2$ first, since we are already familiar with these hypergeometric functions from our study of the $p/q = 4/3$ case. Furthermore, because we need to evaluate ${}_2F_2(1, 3/2; 4/3, 5/3; \pm iz)^{+}$, we do not have to worry about the Stokes phenomenon affecting the L-asymptotics as we found previously. Denoting the L-asymptotics with an L subscript, we obtain

$${}_2F_2\left(1, \frac{3}{2}; \frac{4}{3}, \frac{5}{3}; \pm iz\right)_L^{+} = \frac{\Gamma(4/3)\,\Gamma(5/3)}{\Gamma(3/2)}$$
$$\times\; \left\{\pm z^{-1} e^{i\pi/2} \frac{\Gamma(1/2)}{\Gamma(1/3)\,\Gamma(2/3)}\; {}_3F_1\left(\frac{1}{3}, \frac{2}{3}, 1; \frac{1}{2}; \pm\frac{e^{i\pi/2}}{z}\right)\right.$$
$$\left. (\pm)\; \left(z^{-1} e^{i\pi/2}\right)^{3/2} \frac{\Gamma(3/2)\,\Gamma(7/6)}{\Gamma(1/2)\,\Gamma(1/6)}\; {}_2F_0\left(\frac{5}{6}, \frac{7}{6}; \pm\frac{e^{i\pi/2}}{z}\right)\right\}^{+} \; . \tag{5.71}$$

After some manipulation, Eq. (5.71) becomes

$$
{}_2F_2\left(1,\frac{3}{2};\frac{4}{3},\frac{5}{3};\pm iz\right)_L^+ = \frac{81}{2}\,\Gamma(4/3)\,\Gamma(5/3)\left[\frac{1}{\Gamma(1/3)\,\Gamma(2/3)}\right.
\times \sum_{k=0}^{\infty}\frac{(-1)^{k+1}a^{4k+4}}{(2\pi n)^{6k+6}}\frac{\Gamma(6k+3)}{\Gamma(4k+2)} + \frac{\sqrt{6}\,a^3}{2(2\pi n)^{9/2}}\frac{\Gamma(7/6)}{\Gamma(1/6)}
\times\left(\sum_{k=0}^{\infty}\left(\frac{(-1)^{k+1}a^{4k+4}}{(2\pi n)^{6k}}\frac{\Gamma(6k+3/2)}{\Gamma(4k+1)} + \frac{(-1)^{k+1}a^{4k+2}}{(2\pi n)^{6k+3}}\right.\right.
\left.\left.\left.\times\frac{\Gamma(6k+9/2)}{\Gamma(4k+3)}\right)\right)\right] . \tag{5.72}
$$

In the same manner, we can determine the L-asymptotics for the ${}_1F_1$ hypergeometric function in Eq. (5.58). Hence, we find

$$
G_{2,3}^{1,2}\left(\pm z e^{i\pi/2}\,\middle|\,\begin{matrix}0,1/6\\0,0,1/3\end{matrix}\right)_L^+ = z^{-5/6}\,\frac{\Gamma(5/6)}{\Gamma(-1/6)}\left[e^{-5i\pi/12}\right.
\times\,{}_2F_0\left(\frac{5}{6},\frac{7}{6};\frac{e^{-i\pi/2}}{z}\right) + e^{5i\pi/12}\,{}_2F_0\left(\frac{5}{6},\frac{7}{6};\frac{e^{i\pi/2}}{z}\right)\left.\right]
= \frac{2z^{-5/6}}{3\sqrt{\pi}}\left\{\sum_{k=0}^{\infty}(-1)^{k+1}\left(\frac{4}{27z}\right)^{2k}\left(\frac{\Gamma(6k+3/2)}{\Gamma(4k+1)}\cos\frac{5\pi}{12}\right.\right.
\left.\left.-\frac{4}{27z}\frac{\Gamma(6k+9/2)}{\Gamma(4k+3)}\sin\frac{5\pi}{12}\right)\right\} . \tag{5.73}
$$

The L-asymptotics for the Meijer G-function in Eq. (5.60) are found to be

$$
z^{4/3}G_{2,3}^{1,2}\left(\pm z e^{i\pi/2}\,\middle|\,\begin{matrix}0,-7/6\\0,-4/3,-1\end{matrix}\right)_L^+ = z^{-5/6}e^{-i\pi/2}\frac{\Gamma(7/6)}{\Gamma(1/6)}
\times\left[e^{7i\pi/12}\,{}_2F_0\left(\frac{5}{6},\frac{7}{6};\frac{e^{i\pi/2}}{z}\right) - e^{-7i\pi/12}\,{}_2F_0\left(\frac{5}{6},\frac{7}{6};\frac{e^{-i\pi/2}}{z}\right)\right]
= \frac{z^{-5/6}}{2\sqrt{6}}\left\{\sum_{k=0}^{\infty}(-1)^{k}\left(\frac{4}{27z}\right)^{2k}\left(\frac{4(1-\sqrt{3})}{27z\,\Gamma(3/2)}\frac{\Gamma(6k+9/2)}{\Gamma(4k+3)}\right.\right.
\left.\left.+\frac{(1+\sqrt{3})}{\Gamma(3/2)}\frac{\Gamma(6k+3/2)}{\Gamma(4k+1)}\right)\right\} . \tag{5.74}
$$

To complete the evaluation of the generalised Euler-Jacobi series for $p/q = 3/2$, we must calculate the K-asymptotics for the various hypergeometric functions appearing in Eqs. (5.58)-(5.60). Unlike the (4/3)-case, however, the hypergeometric functions for the (3/2)-case, i.e. Eqs. (5.58)-(5.60), do not have the same number of indices or parameters. As a consequence, two apparently different recursion relations for the K-asymptotics will be obtained by using Eqs. (3.5)-(3.10). Furthermore, we shall show for the (3/2)-case that there are no growing exponentials for the K-asymptotics, and hence, there is no need to be alarmed about a possible divergence as $a \to 0$. This occurs because the recursion relations can be shown to be equivalent and the K-asymptotics cancel completely as in the (4/3)-case. Therefore, not only do specific hypergeometric functions with the same number of indices/parameters possess the same K-asymptotics, except for a power of z and a change in the phase of the trigonometric function, but also higher order hypergeometric functions can possess similar K-asymptotics. Hence, hypergeometric functions can belong to families possessing the same K-asymptotics except for a power of z and a difference in phase in their trigonometric functions.

For ${}_2F_2(1, 3/2; 4/3, 5/3; z)$ we find that $\beta = 1$, $B_1 = 5/2$, $C_1 = 3$ and thus from Eq. (3.4), $\gamma = -1/2$. By using Eq. (3.2), we can write the K-asymptotics for this hypergeometric function as

$$G^{1,2}_{2,3}\left(\pm\, ze^{i\pi/2} \middle| \begin{matrix} 0, -1/2 \\ 0, -2/3, -1/3 \end{matrix}\right)^+_K = \left(\left(ze^{\pm i\pi/2}\right)^{-1/2} \exp\left(ze^{\pm i\pi/2}\right)\right.$$
$$\left. \times \sum_{r=0}^{\infty} N_r \left(ze^{\pm i\pi/2}\right)^{-r}\right)^+ = 2z^{-1/2} \sum_{r=0}^{\infty} N_r\, z^{-r} \cos(z - (2r+1)\pi/4) \ , \qquad (5.75)$$

where the N_r are evaluated via the c_r in Eq. (3.5). Eq. (5.75) can also be written as

$$K_{2,2}\left(ze^{\pm i\pi/2}\right)^+ = 2z^{-1/2} \sum_{k=0}^{\infty} (-1)^k N_{2k}\, z^{-2k} \cos(z - \pi/4)$$
$$+\ 2z^{-1/2} \sum_{k=0}^{\infty} (-1)^k N_{2k+1}\, z^{-(2k+1)} \sin(z - \pi/4) \ . \qquad (5.76)$$

The $T(t)$ and $U(t)$ (see Eqs. (3.9) and (3.10)) for the above-mentioned ${}_2F_2$ become

$$T(t) = (t - 1/2)(t - 1/6)(t + 1/6) \ , \qquad (5.77)$$

and

$$U(t) = (t + 1/2)(t + 1) \ . \qquad (5.78)$$

Thus Eqs. (3.11) and (3.12) become

$$T_{2-s}(s-k) = \sum_{r=0}^{2-s} (-1)^{2-s-r} \frac{(r+s-k-1/2)}{r!\,(2-s-r)!} \left((r+s-k)^2 - 1/36\right) \ , \quad (5.79)$$

and

$$U_{1-s}(s-k) = \sum_{r=0}^{1-s} (-1)^{1-s-r} \frac{(r+s-k+1/2)}{r!\,(1-s-r)!}\, (r+s-k+1) \ . \qquad (5.80)$$

These equations produce the following recursion relation:

$$k\, c_k = \sum_{s=1}^{2} T_{2-s}(s-k)\, c_{k-s} - \sum_{s=1}^{1} U_{1-s}(s-k)\, c_{k-s} \ , \tag{5.81}$$

where $N_k = c_k\left(2,3\,\middle|\,\begin{matrix}1,3/2\\1,4/3,5/3\end{matrix}\right)$. Hence, we find that (1) $N_0 = c_0 = 1$, (2) $N_1 = c_1 = -1/36$ and (3) $N_2 = c_2 = -35/2\,592$. Then we finally get

$$G_{2,3}^{1,2}\left(\pm ze^{i\pi/2}\,\middle|\,\begin{matrix}0,-1/2\\0,-2/3,-1/3\end{matrix}\right)_K^+ = 2z^{-1/2}\cos(z-\pi/4)\Big(1+35/2592z^2 + O(z^{-4})\Big) + 2z^{-1/2}\sin(z-\pi/4)\left(-1/36z+O(z^{-3})\right) \ . \tag{5.82}$$

For the evaluation of the K-asymptotics of Eq. (5.58), we note that $\beta = 1$, $B_1 = 5/6$, $C_1 = 2/3$ and $\gamma = 1/6$. Thus,

$$G_{2,3}^{1,2}\left(\pm ze^{i\pi/2}\,\middle|\,\begin{matrix}0,1/6\\0,0,1/3\end{matrix}\right)_K^+ = 2\sum_{r=0}^{\infty} z^{-(r+1/6)} N_r \cos\left(z-\left(r-\frac{1}{6}\right)\frac{\pi}{2}\right) \ , \tag{5.83}$$

where the N_r are evaluated via the c_r in Eq. (3.5) and are assumed, for the present, to be different from those found in Eq. (5.59). Because q equals unity, there is no contribution from the U_{q-1} in Eq. (3.7) and hence the recursion relation consists of T_{p-1}. For the ${}_1F_1$ in Eq. (5.58), we find that $\omega_0 = 1/6$ and $\omega_1 = -1/6$. With these results, the recursion relation becomes

$$k\, c_k = T_0(1-k)\, c_{k-1} = T(1-k)\, c_{k-1} = (5/6-k)(7/6-k)\, c_{k-1} \ . \tag{5.84}$$

The first three N_r determined from Eq. (5.84) are $N_0 = 1$, $N_1 = c_1 = -1/36$ and $N_2 = c_2 = -35/2592$, and are identical to the N_r found for ${}_2F_2\left(1,3/2;4/3,5/3;\pm ze^{i\pi/2}\right)$. In general, from Eq. (5.84), the c_k are found to be given by

$$c_k = \frac{\Gamma(k+1/6)}{\Gamma(k+1)}\,\frac{\Gamma(k-1/6)}{\Gamma(1/6)\,\Gamma(-1/6)} \ , \tag{5.85}$$

Furthermore, as a consequence of our analysis of $S_{4/3}(a)$, we know that there is another ${}_1F_1$ hypergeometric function, which will possess the K-asymptotics of ${}_1F_1\left(5/6;2/3;\pm ze^{i\pi/2}\right)$ except for a power of z and a difference in phase. This ${}_1F_1$ hypergeometric function will have $\omega_0 = -1/6$ and $\omega_1 = 1/6$ and since $\beta = 1$, we find that $\gamma = -1/6$, $\rho_1 = 4/3$ and $\alpha_1 = 7/6$ by using Eqs. (3.4), (3.13) and (3.15). As expected, the ${}_1F_1$ hypergeometric function is ${}_1F_1\left(7/6;4/3;\pm ze^{i\pi/2}\right)$, which appears in Eq. (5.60). Hence, we get

$$z^{4/3}G_{2,3}^{1,2}\left(\pm ze^{i\pi/2}\,\middle|\,\begin{matrix}0,-7/6\\0,-4/3,-1\end{matrix}\right)_K^+ = 2z^{-1/2}\sum_{k=0}^{\infty}(-1)^k z^{-2k} \times \Big(N_{2k}\cos(z-\pi/4) + z^{-1}N_{2k+1}\sin(z-\pi/4)\Big) \ . \tag{5.86}$$

So far, we have shown that the two ${}_1F_1$ hypergeometric functions for $S_{3/2}(a)$ possess the same K-asymptotics except for a power of z and a difference in phase of the trigonometric functions and that three of the coefficients of the ${}_2F_2$ in $S_{3/2}(a)$ are identical to those evaluated for the ${}_1F_1$ hypergeometric functions. To show that all the coefficients N_r of the ${}_2F_2$ hypergeometric function are equal to their corresponding N_r for the ${}_1F_1$ hypergeometric function, we must prove that the two recursion relations, Eqs. (5.81) and (5.84), are identical.

If we make the substitution $p_k = 2-k$, then the r.h.s. of Eq. (5.84) becomes

$$k\,c_k = \Big((p_k-1)^2 - 1/36\Big)\;c_{k-1}\ , \tag{5.87}$$

whereas Eq. (5.81) can be written as

$$k\,c_k = \Big[(3/2-p_k)\Big((p_k-1)^2 - 1/36\Big) + (p_k - 1/2)\Big((p_k-1)^2 \; - \; 1/36 + p_k - 1\Big)\Big]c_{k-1} + (p_k-1/2)\Big(p_k^2 - 1/36\Big)\,c_{k-2} = \Big((p_k-1)^2 \; - \; 1/36\Big)c_{k-1} + (p_k-1/2)\,(p_k-1)\,c_{k-1} + (p_k-1/2)(p_k^2-1/36)\,c_{k-2}\ . \tag{5.88}$$

The first term on the r.h.s. of Eq. (5.88) is just the recursion relation given in Eq. (5.87); to show that both recursion relations are identical, the remaining terms must equal zero. If we assume that the c_r are given by Eq. (5.87), then $(1-p_k)c_{k-1} = (p_k^2 - 1/36)c_{k-2}$ and after introducing this result in Eq. (5.88), we do indeed find that the remaining terms are equal to zero. Hence, the recursion relations for the ${}_1F_1$ and the ${}_2F_2$ hypergeometric functions in Eqs. (5.58) to (5.60) are identical. This means that all the N_r appearing in the K-asymptotics of these hypergeometric functions are identical.

Now we are able to combine Eqs. (5.76), (5.83) and (5.87) in order to evaluate the K-asymptotics for $S_{3/2}(a)$. After some algebra it can be shown that the total K-asymptotic contribution to $S_{3/2}(a)$ equals zero. Hence, only the L-asymptotics contribute to the result for $S_{3/2}(a)$. Introducing Eqs. (5.72)-(5.74) into Eq. (5.57), we finally get after some algebra

$$S_{3/2}(a) = \frac{2\Gamma(2/3)}{3a^{2/3}} + \frac{1}{2} + 3\sum_{k=0}^{\infty}\frac{(-1)^k a^{4k+2}}{(2\pi)^{6k+4}}\frac{\Gamma(6k+3)}{\Gamma(4k+2)}\,\zeta(6k+4) + \frac{3}{2\sqrt{\pi}}\sum_{k=0}^{\infty}\frac{(-1)^k a^{4k+1}}{(2\pi)^{6k+2}}\left(\frac{\Gamma(6k+3/2)}{\Gamma(4k+1)}\,\zeta(6k+5/2) + \frac{a^2}{(2\pi)^3} \times \frac{\Gamma(6k+9/2)}{\Gamma(4k+3)}\,\zeta(6k+11/2)\right)\ , \tag{5.89}$$

which is precisely the result obtained by Ramanujan and Berndt, Eq. (2.4).

6. INTEGER CASES FOR $S_{p/q}(a)$ WHERE $2 \le p/q \le 7$

In the previous section we showed for various values of p/q less than 2 that the small a-asymptotic series generated from the inversion formula for the generalised Euler-Jacobi series did not contain oscillatory exponential terms and as a consequence, only yielded the zeta series, which had been obtained by Ramanujan and Berndt. In this section we consider integers for $p/q \ge 2$. As a result of considering these p/q values, we evaluate the expected oscillatory exponentials that arise from the non-complete cancellation in the K-asymptotics of $S_{p/q}(a)$. These additional terms in the small a-asymptotics are of particular importance when p/q is an even integer since the zeta series result obtained by Ramanujan and Berndt vanishes for these p/q values.

Although the L-asymptotics vanish for p/q an even integer and have already been evaluated in Sec. 3 for odd integers, we shall see that as p/q increases the evaluation of the K-asymptotics becomes more complicated. Such a result is largely due to the increasing number of terms in Eq. (3.1) as p/q increases, in addition to the greater effort required to evaluate the $\Gamma_p^{1,q}$ in the same equation. This means that more growing exponentials need to be cancelled and furthermore, that more oscillatory decaying exponentials will appear in the final result for small a at these p/q values. Finally, the same problem with the recursion relation for the N_r will appear as it did for $p/q = 3/2$ except that now there will be more hypergeometric functions with greater numbers of parameters. Thus, the recursion relations will become more complicated.

A. Even integers

We begin our analysis of these cases by considering $p/q = 2$. Since $S_2(a)$ has already been discussed in some detail, we shall consider $S_2^r(a)$ before putting $r = 1$ to show that the Meijer G-functions in Eq. (2.16) produce the small a-result for $S_2(a)$ in Eq. (1.2). Putting $p/q = 2$ in Eq. (2.15) yields

$$S_2^r(a) = \frac{\Gamma(r/2)}{2a^{r/2}} + \frac{1}{2}\,\delta_{r,1} + \frac{1}{2}\sqrt{\frac{\pi}{a}}\sum_{n=1}^{\infty}\sum_{l=0}^{1}(-1)^l z^l\, a^{(1-r)/2}$$
$$\times \left[G_{2,3}^{1,2}\left(\pm ze^{i\pi}\,\middle|\,\begin{matrix}0, -(2l+r)/2\\ 0, -(2l+1)/2, -l\end{matrix}\right)^{+} - \frac{(r-1)}{2}\right.$$
$$\left.\times\, G_{2,3}^{1,2}\left(\pm ze^{i\pi}\,\middle|\,\begin{matrix}0, 1-(2l+r)/2\\ 0, -(2l+1)/2, -l\end{matrix}\right)^{+}\right], \tag{6.1}$$

where $z = n^2\pi^2/a$. To evaluate Eq. (6.1), we require the following results:

$$G_{2,3}^{1,2}\left(\pm z e^{i\pi} \,\middle|\, \begin{matrix} 0, -r/2 \\ 0, -1/2, 0 \end{matrix} \right)^{+} = \frac{\Gamma(1+r/2)}{\Gamma(3/2)} \, {}_1F_1\left(1+\frac{r}{2}; \frac{3}{2}; \pm z\right)^{+} , \tag{6.2}$$

$$G_{2,3}^{1,2}\left(\pm z e^{i\pi} \,\middle|\, \begin{matrix} 0, -(r/2+1) \\ 0, -3/2, -1 \end{matrix} \right)^{+} = \frac{\Gamma(1+r/2)}{z\Gamma(3/2)} \, {}_1F_1\left(1+\frac{r}{2}; \frac{3}{2}; \pm z\right)^{-} , \tag{6.3}$$

$$G_{2,3}^{1,2}\left(\pm z e^{i\pi} \,\middle|\, \begin{matrix} 0, 1-r/2 \\ 0, -1/2, 0 \end{matrix} \right)^{+} = \frac{\Gamma(r/2)}{\Gamma(3/2)} \, {}_1F_1\left(\frac{r}{2}; \frac{3}{2}; \pm z\right)^{+} , \tag{6.4}$$

$$G_{2,3}^{1,2}\left(\pm z e^{i\pi} \,\middle|\, \begin{matrix} 0, -r/2 \\ 0, -3/2, -1 \end{matrix} \right)^{+} = \frac{\Gamma(r/2)}{z\Gamma(3/2)} \, {}_1F_1\left(\frac{r}{2}; \frac{3}{2}; \pm z\right)^{-} . \tag{6.5}$$

If we introduce these results into Eq. (6.1), then we obtain

$$S_2^r(a) = \frac{\Gamma(r/2)}{2a^{r/2}} + \frac{1}{2}\,\delta_{r,1} + \frac{\sqrt{\pi}}{a^{r/2}} \sum_{n=1}^{\infty} \left[\frac{\Gamma(1+r/2)}{\Gamma(3/2)} \, {}_1F_1\Big(1+\frac{r}{2}; \frac{3}{2}; -z\Big) - \left(\frac{r-1}{2}\right) \frac{\Gamma(r/2)}{\Gamma(3/2)} \, {}_1F_1\Big(\frac{r}{2}; \frac{3}{2}; -z\Big) \right] . \tag{6.6}$$

Putting $r = 1$ yields the generalised Euler-Jacobi series for $p/q = 2$ as given by Eq. (1.2), since the first ${}_1F_1$ becomes $\exp(-z)$. Furthermore, by utilising No. 7.11.3.4 of Prudnikov et al [22], we can rewrite Eq. (6.6):

$$S_2^r(a) = \frac{\Gamma(r/2)}{2a^{r/2}} + \frac{1}{2}\,\delta_{r,1} + \frac{2^{-r/2-2}}{\sqrt{a}} \sum_{n=1}^{\infty} \frac{e^{-z/2}}{\sqrt{z}} \left[\frac{\Gamma(1+r/2)}{\Gamma(3/2)} \, \Gamma(-r/2) \times \Big(D_r\big(-\sqrt{2z}\big) - D_r\big(\sqrt{2z}\big)\Big) - (r-1)\, \frac{\Gamma(r/2)}{\Gamma(3/2)} \, \Gamma(1-r/2) \times \Big(D_{r-2}\big(-\sqrt{2z}\big) - D_{r-2}\big(\sqrt{2z}\big)\Big) \right] , \tag{6.7}$$

where $D_r(z)$ represents a parabolic cylinder function. Specific results can also be obtained by setting r equal to integers between 2 and 6 and using the table of values for ${}_1F_1$ on p. 580 of Ref. [22]. For example, one finds that

$$S_2^5(a) = \sum_{n=1}^{\infty} e^{-an^2} n^4 = \frac{\Gamma(5/2)}{2a^{5/2}} + \frac{1}{a^2}\sqrt{\frac{\pi}{a}} \sum_{n=1}^{\infty} \big(3/4 - 3z + z^2\big) e^{-z} , \tag{6.8}$$

and

$$S_2^2(a) = \sum_{n=1}^{\infty} e^{-an^2} n = \frac{1}{2a} + \frac{1}{a} \sum_{n=1}^{\infty} \left[1 - \sqrt{\pi z}\, e^{-z} erfi\big(\sqrt{z}\big) \right] , \tag{6.9}$$

where in Eq. (6.9) we have used the result concerning the error function on p. 725 of Ref. [25]. Eq. (6.8) can also be obtained by evaluating the double derivative of Eq. (1.2) w.r.t. a. It should be noted that we have taken the positive root of $\sqrt{-1}$ in obtaining Eq. (6.9).

To obtain an asymptotic expansion for $S_2^r(a)$, we can use Luke's asymptotic expansion [8] of ${}_1F_1(z)$ for large z. For $\delta - \pi/2 \le \arg z \le 3\pi/2 - \delta$, we obtain

$$\frac{\Gamma(1+r/2)}{\Gamma(3/2)}\,{}_1F_1\Big(1+\frac{r}{2};\frac{3}{2};-z\Big) \sim e^{-z}\left(ze^{-i\pi}\right)^{(r-1)/2}\sum_{k=0}^{\infty}(-1)^k N_k\, z^{-k} + z^{-(1+r/2)}\frac{\Gamma(1+r/2)}{\Gamma((1-r)/2)}\,{}_2F_0\Big(1+\frac{r}{2},\frac{(r+1)}{2};\frac{1}{z}\Big)\ , \tag{6.10}$$

whereas for $\delta - 3\pi/2 \le \arg z \le \pi/2 - \delta$, we get

$$\frac{\Gamma(1+r/2)}{\Gamma(3/2)}\,{}_1F_1\Big(1+\frac{r}{2};\frac{3}{2};-z\Big) \sim e^{-z}\left(ze^{i\pi}\right)^{(r-1)/2}\sum_{k=0}^{\infty}(-1)^k N_k\, z^{-k} + z^{-(1+r/2)}\frac{\Gamma(1+r/2)}{\Gamma((1-r)/2)}\,{}_2F_0\Big(1+\frac{r}{2},\frac{(r+1)}{2};\frac{1}{z}\Big)\ . \tag{6.11}$$

Similar expressions can be obtained for the other hypergeometric function in Eq. (6.6) by replacing $1+r/2$ with $r/2$ in Eqs. (6.10) and (6.11). It should also be noted that the recursion relation for the N_k is to be modified accordingly.

The apparent discrepancy between Eqs. (6.10) and (6.11) in the region where $|\arg z| < \pi/2$ is attributed to the Stokes phenomenon. In this strip, we notice that if $r-1 \ne 2m$ where $m \in \mathbf{Z}$, then both Eqs. (6.10) and (6.11) will become complex and different due to the phase of $\exp(\pm i\pi(r-1)/2)$. We note at this point that a similar type of crossover to complex behaviour also arises in Berry's work [20] on the smoothing of Stokes' discontinuities. While this effect does not worry us here, for the complex term is subdominant to the leading real term, we do, nonetheless, feel that this is an area of asymptotics that could be quite interesting in itself and worthy of future study. However, since we are only interested in large z, the expression involving the ${}_2F_0$ dominates in both Eqs. (6.10) and (6.11). Hence, the ${}_1F_1$ hypergeometric functions in Eq. (6.6) can be approximated by

$$\frac{\Gamma(1+r/2)}{\Gamma(3/2)}\,{}_1F_1\Big(1+\frac{r}{2};\frac{3}{2};-z\Big) \sim \frac{\Gamma(1+r/2)}{\Gamma((1-r)/2)}\,z^{-(1+r/2)} \times\left(1+\frac{(1+r/2)}{2z}(r+1)+O\big(z^{-2}\big)\right)\ , \tag{6.12}$$

and

$$\frac{\Gamma(r/2)}{\Gamma(3/2)}\,{}_1F_1\Big(\frac{r}{2};\frac{3}{2};-z\Big) \sim \frac{\Gamma(r/2)}{\Gamma((3-r)/2)}\,z^{-r/2} \times\left(1+\frac{r}{4z}(r-1)+O\big(z^{-2}\big)\right)\ . \tag{6.13}$$

The leading terms in Eqs. (6.12) and (6.13) correspond to the result for $\phi = \pi$ obtained on p. 611 of Morse and Feshbach [19]. Introducing Eqs. (6.12) and (6.13) into Eq. (6.6), we get

$$S_2^r(a) \sim \frac{\Gamma(r/2)}{2a^{r/2}} + \frac{1}{2}\,\delta_{r,1} + \frac{\Gamma(r/2)\,\pi^{1/2-r}}{\Gamma((1-r)/2)} \Big(\zeta(r) + \frac{a(r^2+r)}{4\pi^2}\,\zeta(r+2) + O\big(a^2\big)\Big) \ . \tag{6.14}$$

To complete our study of the $p/q = 2$ case, we note that the m-fold derivative of Eq. (6.14) w.r.t. r yields the small a-expansion for $S_2^{r,m}(a)$. In particular, for $m = 1$ the following result is obtained:

$$S_2^{r,1}(a) = \sum_{n=1}^{\infty} e^{-an^2} n^{r-1} \log n \sim \frac{\Gamma(r/2)}{4a^{r/2}}\,(\Psi(r/2) - \log a)$$
$$+ \frac{\Gamma(r/2)\,\pi^{1/2-r}}{2\Gamma((1-r)/2)}\Big(\Psi(r/2) - 2\log\pi + \Psi((1-r)/2)\Big)\Big(\zeta(r)$$
$$+ \frac{a(r^2+r)}{4\pi^2}\,\zeta(r+2)\Big) + \frac{\Gamma(r/2)\pi^{1/2-r}}{\Gamma((1-r)/2)}\Big(\zeta'(r) + \frac{(2r+1)a}{4\pi^2}\,\zeta(r+2)$$
$$+ \frac{a(r^2+r)}{4\pi^2}\,\zeta'(r+2)\Big) + O\big(a^2\big) \ . \tag{6.15}$$

We now consider the $p/q = 4$ case for the generalised Euler-Jacobi series. Inserting $p/q = 4$ in Eq. (2.16) yields

$$S_4(a) = \frac{\Gamma(1/4)}{4a^{1/4}} + \frac{1}{2} + \frac{(2\pi)^{3/2}}{8a^{1/4}} \sum_{n=1}^{\infty}\sum_{l=0}^{3} (-1)^l z^{l/2} \times G_{1,4}^{1,1}\left(\pm z \,\middle|\, \begin{matrix} 0 \\ 0, -l/2, 1/4 - l/2, (1-l)/2 \end{matrix}\right)^{+} , \tag{6.16}$$

where $z = (n\pi/2)^4 a^{-1}$. An alternative representation for $S_4(a)$ can also be obtained from Eq. (2.18), which is

$$S_4(a) = \frac{\Gamma(1/4)}{4a^{1/4}} + \frac{1}{2} + \frac{(2\pi)^{3/2}}{8a^{1/4}} \sum_{n=1}^{\infty}\sum_{l=0}^{3} (-1)^l z^{l/2} \times G_{4,1}^{1,1}\left(\pm z \,\middle|\, \begin{matrix} 1, (2l+2)/4, (2l+3)/4, l/2+1 \\ (2l+1)/4 \end{matrix}\right)^{+} . \tag{6.17}$$

To evaluate these expressions, we require

$$G_{1,4}^{1,1}\left(\pm z \,\middle|\, \begin{matrix} 0 \\ 0, 0, 1/4, 1/2 \end{matrix}\right)^{+} = G_{4,1}^{1,1}\left(\pm z^{-1} \,\middle|\, \begin{matrix} 1, 1/2, 3/4, 1 \\ 1/4 \end{matrix}\right)^{+} = \frac{{}_0F_2(1/2, 3/4; \pm z)^{+}}{\Gamma(1/2)\,\Gamma(3/4)} , \tag{6.18}$$

$$G_{1,4}^{1,1}\left(\pm z \,\middle|\, \begin{matrix} 0 \\ 0, -1/2, -1/4, 0 \end{matrix}\right)^{+} = G_{4,1}^{1,1}\left(\pm z^{-1} \,\middle|\, \begin{matrix} 1, 1, 5/4, 3/2 \\ 3/4 \end{matrix}\right)^{+} = \frac{{}_0F_2(5/4, 3/2; \pm z)^{+}}{\Gamma(3/2)\,\Gamma(5/4)} , \tag{6.19}$$

$$G_{1,4}^{1,1}\left(\pm z \,\middle|\, \begin{matrix} 0 \\ 0,-1,-3/4,-1/2 \end{matrix}\right)^{+} = G_{4,1}^{1,1}\left(\pm z^{-1} \,\middle|\, \begin{matrix} 1,1,5/4,6/4 \\ 5/4 \end{matrix}\right)^{+}$$
$$= \frac{{}_0F_2(1/2,3/4;\pm z)^{-}}{z\Gamma(1/2)\,\Gamma(3/4)} , \qquad (6.20)$$

$$G_{1,4}^{1,1}\left(\pm z \,\middle|\, \begin{matrix} 0 \\ 0,-3/2,-5/4,-1 \end{matrix}\right)^{+} = G_{4,1}^{1,1}\left(\pm z^{-1} \,\middle|\, \begin{matrix} 1,2,9/4,5/2 \\ 7/4 \end{matrix}\right)^{+}$$
$$= \frac{{}_0F_2(3/2,5/4;\pm z)^{-}}{z\Gamma(3/2)\,\Gamma(5/4)} . \qquad (6.21)$$

Eqs. (6.18) to (6.21) confirm the observation made in Sec. 3 that for even integers the generalised Euler-Jacobi series will only consist of ${}_0F_p$ hypergeometric functions. Hence, there will be no L-asymptotic contribution to the final result for $S_4(a)$. As a consequence, we do not expect a zeta series, but should obtain a series consisting of oscillatory decaying exponentials with the leading term equal to Eq. (4.16).

After introducing Eqs. (6.18)-(6.21) into Eq. (6.16), we arrive at

$$S_4(a) = \frac{\Gamma(1/4)}{4a^{1/4}} + \frac{1}{2} + \frac{\pi^{3/2}}{\sqrt{2}\,a^{1/4}} \sum_{n=1}^{\infty} \Big[\big(\Gamma(1/2)\,\Gamma(3/4)\big)^{-1} {}_0F_2(1/2,3/4;z)$$
$$- \sqrt{z}\,\big(\Gamma(5/4)\,\Gamma(3/2)\big)^{-1} {}_0F_2(5/4,3/2;z)\Big] . \qquad (6.22)$$

In applying Eq. (3.1) to Eq. (6.22), we note that r can either equal 1 so that $\delta - 2\pi \le \arg z \le 3\pi/2 - \delta$, or it can equal 2 so that $\delta - 3\pi/2 \le \arg z \le 2\pi - \delta$. We also find that the $T(t)$ as given by Eq. (3.9) are identical for both hypergeometric functions so that Eq. (3.2) for both hypergeometric functions becomes

$$K_{0,2}(z) = \frac{1}{2\sqrt{3}\,\pi}\, e^{3z^{1/3}} z^{\gamma} \left(1 + \sum_{k=1}^{\infty} N_k\, z^{-k/3}\right) , \qquad (6.23)$$

where the N_k are the same recursion relation

$$3k\, c_k = (T(2-k) - T(1-k))\, c_{k-1} + T(2-k)\, c_{k-2} , \qquad (6.24)$$

with $3^{-k} c_k = N_k$. In Eq. (6.24), the $T(t)$ are given by

$$T(t) = (t - 1/4)(t - 7/4)(t - 1) . \qquad (6.25)$$

In addition, for ${}_0F_2(1/2,3/4;z)$, $\gamma = -1/12$ whereas for ${}_0F_2(5/4,3/2;z)$, $\gamma = -7/12$.

If r is equal to 1, then we shall require $\bar{\Gamma}_3^{1,0}(1)$ in order to evaluate the second sum in Eq. (3.1). On the other hand, if r equals 2 then we shall need

$\Gamma_3^{1,0}(1)$ for the evaluation of the first sum in the same equation. According to Luke [8], $\Gamma_{q+1}^{1,p}(1)$ for the hypergeometric function ${}_pF_q(\alpha_1, \cdots, \alpha_p; \beta_1, \cdots, \beta_q; z)$ is given by

$$\Gamma_{q+1}^{1,p}(1) = \sum_{j=1}^{p} \exp(-2\pi i \alpha_j) - \sum_{j=1}^{q} \exp(-2\pi i \beta_j) \ , \tag{6.26}$$

while $\bar{\Gamma}_{q+1}^{1,p}(1)$ is the complex conjugate of $\Gamma_{q+1}^{1,p}(1)$. Thus for ${}_0F_2(1/2, 3/4; z)$ and ${}_0F_2(5/4, 3/2; z)$, the $\Gamma_3^{1,0}(1)$ are, respectively, equal to $1 - i$ and $1 + i$.

For each value of r, two different forms for each ${}_0F_2$ appearing in Eq. (6.22) can be obtained from Eq. (3.1), depending on whether we choose $z = -ze^{i\pi}$ or $z = -ze^{-i\pi}$, since $\delta - 3\pi/2 \leq \arg(ze^{\pm i\pi}) \leq 2\pi - \delta$. For $r = 2$ with z equal to $-ze^{i\pi}$, we find that by setting $y = 3z^{1/3}$,

$$\begin{aligned} {}_0F_2(\rho_1, \rho_2; z) &\sim \frac{\Gamma(\rho_1)\,\Gamma(\rho_2)}{2\sqrt{3}\,\pi}\, z^{\gamma} \sum_{k=0}^{\infty} N_k\, z^{-k/3} \Big\{ e^{y} + e^{-y/2} \\ &\times \Big(i\epsilon\, e^{-i\sqrt{3}y/2 + 2i\pi k/3 - 2\pi i\gamma} + 2\cos\big(\sqrt{3}\,y/2 - 2\pi k/3 + 2\pi\gamma\big)\Big)\Big\} \ , \end{aligned} \tag{6.27}$$

whereas for $r = 1$ and $z = -ze^{i\pi}$, we get

$$\begin{aligned} {}_0F_2(\rho_1, \rho_2; z) &\sim \frac{\Gamma(\rho_1)\,\Gamma(\rho_2)}{2\sqrt{3}\,\pi}\, z^{\gamma} \sum_{k=0}^{\infty} N_k\, z^{-k/3} \Big\{ e^{y} + e^{3\pi i\gamma - y/2} \\ &\times \Big(-i\epsilon e^{-i\sqrt{3}y/2 + 2i\pi k/3 + \pi i\gamma} + 2\cos\big(\sqrt{3}\,y/2 - 2\pi k/3 - \pi\gamma\big)\Big)\Big\} \ , \end{aligned} \tag{6.28}$$

where it has been assumed that $\Gamma_3^{1,0}(1)$ is equal to $1 + \epsilon i$ and ϵ is -1 or 1 respectively. Otherwise one would have to return to Eq. (3.1) and re-evaluate the K-asymptotics accordingly. The K-asymptotics for ${}_0F_2(\rho_1, \rho_2; z)$ with $r = 1$ and $z = -ze^{-i\pi}$ are equal to the complex conjugate of Eq. (6.27) while those for ${}_0F_2(\rho_1, \rho_2; z)$ with $r = 2$ and $z = -ze^{-i\pi}$ are equal to the complex conjugate of Eq. (6.28).

Once again, we note the crossover to complex behaviour in the various expansions for ${}_0F_2(\rho_1, \rho_2; z)$ as occurred for $p/q = 2$. Normally in dealing with these expansions one would neglect the complex terms since these are subdominant to the leading exponential e^y. However, as we noted in Sec. 3, the leading exponential is a growing term and since the generalised Euler-Jacobi series is convergent, all leading exponential terms must cancel, thereby leaving the complex subdominant terms. We must, therefore, look for methods of combining these complex terms into a real entity. Our task is made simpler since we know from our steepest descent analysis in Sec. 4 that the leading term for $S_4(a)$ is given by Eq. (4.16).

If we now introduce the appropriate values for γ, ρ_1 and ρ_2, then we obtain the following result using the ${}_0F_2$ asymptotic expansion given by Eq. (6.27):

$$\left[\left(\Gamma\Big(\frac{1}{2}\Big)\,\Gamma\Big(\frac{3}{4}\Big)\right)^{-1} {}_0F_2\Big(\frac{1}{2}, \frac{3}{4}; z\Big) - \sqrt{z}\,\left(\Gamma\Big(\frac{5}{4}\Big)\,\Gamma\Big(\frac{3}{2}\Big)\right)^{-1} {}_0F_2\Big(\frac{5}{4}, \frac{3}{2}; z\Big) \right]$$

$$\sim \frac{2z^{-1/12}}{\sqrt{3}\pi} \sum_{k=0}^{\infty} N_k \, z^{-k/3} e^{-y/2} \cos\left(\frac{\sqrt{3}}{2} y - \frac{k\pi}{3} - \frac{\pi}{6}\right) . \tag{6.29}$$

If instead Eq. (6.28) is utilised, the same result ensues. Furthermore, since the remaining cases are complex conjugates of the two previous cases, Eq. (6.29) holds for these ${}_0F_2$ asymptotic expansions. Thus introducing Eq. (6.29) into Eq. (6.22) yields

$$S_4(a) \sim \frac{\Gamma(1/4)}{4a^{1/4}} + \frac{1}{2} + \frac{2^{5/6}\pi^{1/6}}{\sqrt{3}\, a^{1/6}} \sum_{n=1}^{\infty} \sum_{k=0}^{\infty} \frac{N_k}{n^{1/3}} \, z^{-k/3} \, e^{-3z^{1/3}/2} \times \cos\left(\frac{3\sqrt{3}}{2} \, z^{1/3} - \frac{k\pi}{3} - \frac{\pi}{6}\right) , \tag{6.30}$$

where $N_0 = 1$, $N_1 = 7/144$, $N_2 = 385/41\,472$, $N_3 = 39\,655/17\,915\,904$, $N_4 = 665\,665/10\,319\,560\,704$ and the other values up to N_{20} are presented in Table 1. These N_k have been determined after rewriting Eq. (6.24) as

$$c_k = (7/48k - 1 + k) \, c_{k-1} + (7/48k - 13/16 + k - k^2/3) \, c_{k-2} \quad . \tag{6.31}$$

Note that the leading term in Eq. (6.30), i.e. $N_0 = 1$, is identical to the result obtained by the method of steepest descent (Eq. (4.16)).

The derivation of Eq. (6.30) using our generalised Euler-Jacobi inversion formula given by either Eq. (2.14) or Eq. (2.16) is quite remarkable since we have had to rely on asymptotic expansions of ${}_0F_2$, which were determined using Luke's prescription [8]. Normally, one would neglect the subdominant complex terms in these expansions, which emerge as a consequence of the Stokes phenomenon. However, what we have found in evaluating $S_4(a)$ is that the leading exponentials cancel each other and that the subdominant complex terms combine to form the real entity given by Eq. (6.30). Furthermore, although four separate forms for the ${}_0F_2$ hypergeometric functions exist, i.e. Eqs. (6.27), (6.28) and their complex conjugates, each of the forms gives Eq. (6.30) when introduced into Eq. (6.22).

To complete this section, we consider the $p = 6$ case to demonstrate the utility of our inversion formula over the method of steepest descent. According to Eq. (2.16), the generalised Euler-Jacobi series for $p = 6$ can be written as

$$S_6(a) = \frac{\Gamma(1/6)}{6a^{1/6}} + \frac{1}{2} + \frac{(2\pi)^{5/2}}{6\sqrt{6}\, a^{1/6}} \sum_{n=1}^{\infty} \sum_{l=0}^{5} (-1)^l z^{l/3} \times G_{1,6}^{1,1}\left(\pm z \,\middle|\, \begin{matrix} 0 \\ 0, -l/3, 1/6 - l/3, 1/3 - l/3, 1/2 - l/3, 2/3 - l/3 \end{matrix}\right)^{+} , \tag{6.32}$$

where z now equals $(n\pi/3)^6 a^{-1}$. To evaluate Eq. (6.32), we require

$$G_{1,6}^{1,1}\left(\pm z \left| \begin{matrix} 0 \\ 0, 0, 1/6, 1/3, 1/2, 2/3 \end{matrix} \right.\right)^+ = \Big(\Gamma(1/3)\,\Gamma(1/2)\Big)^{-1}$$
$$\times\ \Big(\Gamma(2/3)\,\Gamma(5/6)\Big)^{-1}\,{}_0F_4\Big(1/3, 1/2, 2/3, 5/6; \pm z\Big)^+ , \tag{6.33}$$

$$G_{1,6}^{1,1}\left(\pm z \left| \begin{matrix} 0 \\ 0, -1/3, -1/6, 0, 1/6, 1/3 \end{matrix} \right.\right)^+ = \Big(\Gamma(2/3)\,\Gamma(5/6)\Big)^{-1}$$
$$\times\ \Big(\Gamma(7/6)\,\Gamma(4/3)\Big)^{-1}\,{}_0F_4\Big(2/3, 5/6, 7/6, 4/3; \pm z\Big)^+ , \tag{6.34}$$

$$G_{1,6}^{1,1}\left(\pm z \left| \begin{matrix} 0 \\ 0, -2/3, -1/2, -1/3, -1/6, 0 \end{matrix} \right.\right)^+ = \Big(\Gamma(7/6)\,\Gamma(4/3)\Big)^{-1}$$
$$\times\ \Big(\Gamma(3/2)\,\Gamma(5/3)\Big)^{-1}\,{}_0F_4\Big(7/6, 4/3, 3/2, 5/3; \pm z\Big)^+ , \tag{6.35}$$

$$G_{1,6}^{1,1}\left(\pm z \left| \begin{matrix} 0 \\ 0, -1, -5/6, -2/3, -1/2, -1/3 \end{matrix} \right.\right)^+ = \Big(z\Gamma(1/3)\Big)^{-1}$$
$$\times\ \Big(\Gamma(1/2)\,\Gamma(2/3)\,\Gamma(5/6)\Big)^{-1}\,{}_0F_4\Big(1/3, 1/2, 2/3, 5/6; \pm z\Big)^- , \tag{6.36}$$

$$G_{1,6}^{1,1}\left(\pm z \left| \begin{matrix} 0 \\ 0, -4/3, -7/6 - 1, -5/6 - 2/3 \end{matrix} \right.\right)^+ = \Big(z\Gamma(2/3)\Big)^{-1}$$
$$\times\ \Big(\Gamma(5/6)\,\Gamma(7/6)\,\Gamma(4/3)\Big)^{-1}\,{}_0F_4\Big(2/3, 5/6, 7/6, 4/3; \pm z\Big)^- , \tag{6.37}$$

$$G_{1,6}^{1,1}\left(\pm z \left| \begin{matrix} 0 \\ 0, -5/3, -3/2, -4/3, -7/6, -1 \end{matrix} \right.\right)^+ = \Big(z\Gamma(7/6)\Big)^{-1}$$
$$\times\ \Big(\Gamma(4/3)\,\Gamma(3/2)\,\Gamma(5/3)\Big)^{-1}\,{}_0F_4\Big(7/6, 4/3, 3/2, 5/3; \pm z\Big)^- . \tag{6.38}$$

If Eqs. (6.33) to (6.38) are introduced into Eq. (6.32), then we find

$$S_6(a) = \frac{\Gamma(1/6)}{6a^{1/6}} + \frac{1}{2} + \frac{2(2\pi)^{5/2}}{6\sqrt{6}a^{1/6}} \sum_{n=1}^{\infty} \left[\frac{{}_0F_4(1/3, 1/2, 2/3, 5/6; -z)}{\Gamma(1/3)\,\Gamma(1/2)\,\Gamma(2/3)\,\Gamma(5/6)} \right.$$
$$\left. - \frac{z^{1/3}\,{}_0F_4\big(2/3, 5/6, 7/6, 4/3; -z\big)}{\Gamma(2/3)\,\Gamma(5/6)\,\Gamma(7/6)\,\Gamma(4/3)} + \frac{z^{2/3}\,{}_0F_4(7/6, 4/3, 3/2, 5/3; -z)}{\Gamma(7/6)\,\Gamma(4/3)\,\Gamma(3/2)\,\Gamma(5/3)} \right] . \tag{6.39}$$

The above equation confirms the result found in Sec. 3 that there are no L-asymptotics for $p = 6$. Also according to Sec. 3, each of the ${}_0F_4$ hypergeometric functions in Eq. (6.39) has a growing exponential of the form of $\exp(5z^{1/5}\cos(\pi/5))$, which must cancel. Hence, all the ${}_0F_4$ hypergeometric functions possess similar K-asymptotics, so we need only evaluate the T-function for the recursion relation using one of the ${}_0F_4$ hypergeometric functions. This is found to be

$$T(t) = (t-1/3)(t-17/6)(t-11/3)(t-2)(t-7/6) \;, \tag{6.40}$$

which is then to be employed in the recursion relation

$$5k\, c_k = \sum_{s=1}^{4} T_{4-s}(s-k)\, c_{k-s} \;, \tag{6.41}$$

with $5^{-k}c_k = N_k$. Eq. (6.41) eventually yields

$$\begin{aligned} 5k\, c_k &= (55/36 - 10k + 10k^2)\, c_{k-1} + 5(11/12 - 59k/12 + 6k^2 \\ &- 2k^3)\, c_{k-2} + (1397/162 - 175k/4 + 715k^2/12 - 30k^3 + 5k^4)\, c_{k-3} \\ &+ (2618/324 - 13045k/324 + 19170k^3/324 + 10k^4 - k^5)\, c_{k-4} \;. \end{aligned} \tag{6.42}$$

If we put r equal to 2 in Eq. (3.1), then each ${}_0F_4(\rho_1, \rho_2, \rho_3, \rho_4; -z)$ in Eq. (6.39) can be expressed asymptotically as

$$\begin{aligned} {}_0F_4(\rho_1,\rho_2,\rho_3,\rho_4;-z) &\sim \Gamma(\rho_1)\,\Gamma(\rho_2)\,\Gamma(\rho_3)\,\Gamma(\rho_4)\,\Big\{K_{0,4}\Big(ze^{\pm i\pi}\Big)^{+} \\ &+ \Gamma_5^{1,0}(1)K_{0,4}\Big(ze^{-3i\pi}\Big) + \bar{\Gamma}_5^{1,0}(1)K_{0,4}\Big(ze^{3i\pi}\Big) + \Gamma_5^{1,0}(2)K_{0,4}\Big(ze^{-5i\pi}\Big)\Big\} \;, \end{aligned} \tag{6.43}$$

where $\Gamma_5^{1,0}(2)$ is given by

$$\Gamma_5^{1,0}(2) = \sum_{1\leq j<k\leq 4} e^{-2i\pi(\rho_j+\rho_k)} \;. \tag{6.44}$$

Full details regarding the evaluation of $\Gamma_5^{1,0}(2)$ can be found on p. 196 of Luke [8].

For ${}_0F_4(1/3,1/2,2/3,5/6;-z)$, we find that $\gamma = -1/15$, $\Gamma_5^{1,0}(0) = 1$, $\Gamma_5^{1,0}(1) = \sqrt{3}\exp(-i\pi/6)$, and $\Gamma_5^{1,0}(2) = 2\exp(-i\pi/3)$. Then using Eq. (6.43) we obtain

$$\begin{aligned} \frac{{}_0F_4(1/3,1/2,2/3,5/6;-z)}{\Gamma(1/3)\,\Gamma(1/2)\,\Gamma(2/3)\,\Gamma(5/6)} &\sim \frac{z^{-1/15}}{2\sqrt{5}\,\pi^2}\sum_{k=0}^{\infty} N_k\, z^{-k/5}\Big[e^{5z^{1/5}\cos(\pi/5)} \\ &\times \cos\Big(5z^{1/5}\sin(\pi/5) - (k+1/3)\pi/5\Big) + \sqrt{3}\, e^{5z^{1/5}\cos(3\pi/5)} \\ &\times \cos\Big(5z^{1/5}\sin(3\pi/5) - (3k+1/6)\pi/5\Big) + (-1)^k e^{-5z^{1/5}}\Big] \;. \end{aligned} \tag{6.45}$$

For ${}_0F_4(2/3,5/6,7/6,4/3;-z)$, $\gamma = -2/5$, $\Gamma_5^{1,0}(0) = 1$, $\Gamma_5^{1,0}(1) = 0$ and $\Gamma_5^{1,0}(2) = 1$. Hence, the K-asymptotics for this hypergeometric function can be written as

$$\frac{-z^{1/3}\,{}_0F_4(2/3,5/6,7/6,4/3;-z)}{\Gamma(2/3)\,\Gamma(5/6)\,\Gamma(7/6)\,\Gamma(4/3)} \sim \frac{-z^{-1/15}}{4\sqrt{5}\,\pi^2} \sum_{k=0}^{\infty} N_k\, z^{-k/5}$$
$$\times \left[2e^{5z^{1/5}\cos(\pi/5)} \cos\Big(5z^{1/5}\sin(\pi/5) - (k+2)\pi/5\Big) + (-1)^k e^{-5z^{1/5}}\right] . \quad (6.46)$$

Finally, for ${}_0F_4(7/6,4/3,3/2,5/3;-z)$, $\gamma = -11/15$, $\Gamma_5^{1,0}(0) = 1$, $\Gamma_5^{1,0}(1) = \sqrt{3}\exp(i\pi/6)$ and $\Gamma_5^{1,0}(2) = 2\exp(i\pi/3)$. Thus the final term in Eq. (6.39) becomes

$$\frac{z^{2/3}\,{}_0F_4(7/6,4/3,3/2,5/3;-z)}{\Gamma(7/6)\,\Gamma(4/3)\,\Gamma(3/2)\,\Gamma(5/3)} \sim \frac{z^{-1/15}}{2\sqrt{5}\,\pi^2} \sum_{k=0}^{\infty} N_k\, z^{-k/5} \Big[e^{5z^{1/5}\cos(\pi/5)}$$
$$\times \cos\Big(5z^{1/5}\sin(\pi/5) - (k+11/3)\pi/5\Big) + \sqrt{3}\, e^{5z^{1/5}\cos(3\pi/5)}$$
$$\times \cos\Big(5z^{1/5}\sin(3\pi/5) - (3k+11/6)\pi/5\Big) + (-1)^k e^{-5z^{1/5}}\Big] . \quad (6.47)$$

In marked contrast to the $p = 4$ case, we note that there are no complex terms appearing in Eqs. (6.45) to (6.47), primarily because the variable in each hypergeometric function is $-z$ and not z as for $p = 4$.

If we introduce Eqs. (6.45) to (6.47) into Eq. (6.39), then we find that all the terms involving $\exp(5z^{1/5}\cos(\pi/5))$ cancel each other as expected, and we are left with

$$S_6(a) \sim \frac{\Gamma(1/6)}{6a^{1/6}} + \frac{1}{2} + \sqrt{\frac{\pi}{15}}\,\frac{1}{a^{1/6}} \sum_{n=1}^{\infty}\sum_{k=0}^{\infty} N_k\, z^{-(k+1/3)/5} \Big[2e^{5z^{1/5}\cos(3\pi/5)}$$
$$\times \cos\Big(5z^{1/5}\sin(3\pi/5) - (3k+1)\pi/5\Big) + (-1)^k e^{-5z^{1/5}}\Big] , \quad (6.48)$$

where $N_0 = 1$, $N_1 = 11/180$, $N_2 = 517/64\,800$, $N_3 = -22\,253/174\,960\,000$ and the remaining N_k can be determined from Eq. (6.42). It can also be seen that unlike the $p=2$ and $p=4$ cases, there is an additional exponential series, which has arisen because $(2m+1)\pi/\beta$ ($m \in \mathbf{Z}^+$) is greater than $\pi/2$ and less than or equal to π for two values of k since $\beta = 5$ for the hypergeometric functions appearing in Eq. (6.39). We should, therefore, expect that as p or β increases, there should be more decaying exponentials appearing in the asymptotic results for generalised Euler-Jacobi series. This becomes even more conspicuous when we study the $p = 3$, 5 and 7 cases in the next subsection.

B. Odd integers

As we shall see, the asymptotic forms for the odd integer cases of $S_p(a)$ are considerably more difficult to evaluate than those for the even integer cases. This is not only due to the fact that there is a contribution from the

L-asymptotics in each $S_{2k+1}(a)$ (see Sec. 3), which yields the Ramanujan-Berndt result Eq. (2.2), but also to the fact that many of the Meijer G-functions or hypergeometric functions appearing in our inversion formula do not reduce to simpler forms as we found for even integers. For example, we found that $S_6(a)$ could only be expressed in terms of three ${}_0F_4$ hypergeometric functions whose gamma functions were relatively simple to evaluate. However, we shall see that $S_7(a)$ consists of seven higher order hypergeometric functions, whose gamma functions will require considerably more attention than those for $S_6(a)$. In addition, the evaluation of the recursion relation requires much greater effort. Nevertheless, for $p = 5$ and 7, we shall introduce techniques which will facilitate the task of evaluating the asymptotics.

We begin with $p = 3$, which is the case that motivated this entire work. Before using our generalised Euler-Jacobi inversion formula, we return to Eq. (2.6) and note that

$$S_3^r(a) = \frac{\Gamma(r/3)}{3a^{r/3}} + \frac{1}{2}\,\delta_{r,1} + \frac{1}{\pi}\sum_{n=1}^{\infty}\frac{1}{n}\int_0^{\infty} dy\, e^{-ay^3}\sin\beta y \times \left(3ay^{r+1} - (r-1)y^{r-2}\right) \;, \tag{6.49}$$

where we have made the change of variable $y = t^{1/3}$. For $r - 2 = 2k$, where k is a natural number, we can use No. 2.5.38.15 from Prudnikov et al [23] to obtain

$$S_3^{2k+2}(a) = \frac{\Gamma((2k+2)/3)}{3a^{(2k+2)/3}} + \frac{1}{\pi}\sum_{n=1}^{\infty}\frac{1}{n}\Bigg[3(-1)^k a\frac{\partial^{2k+3}}{\partial\beta^{2k+3}}\Bigg\{\frac{z}{2\beta}\Bigg(e^{\pi i/4} \times S_{0,1/3}\Big(ze^{3\pi i/4}\Big) + e^{-\pi i/4}S_{0,1/3}\Big(ze^{-3\pi i/4}\Big)\Bigg)\Bigg\} - (-1)^k i(2k+1) \times \frac{\partial^{2k}}{\partial\beta^{2k}}\Bigg\{\frac{z}{2\beta}\Bigg(e^{\pi i/4}S_{0,1/3}\Big(ze^{3\pi i/4}\Big) - e^{-\pi i/4}S_{0,1/3}\Big(ze^{-3\pi i/4}\Big)\Bigg)\Bigg\}\Bigg] \;, \tag{6.50}$$

where $S_{0,1/3}(z)$ is a Lommel function, $z = 2(\beta/3)^{3/2}a^{-1/2}$ and β is to be replaced by $2n\pi$ after the differentiations have been effected. For the case $r - 2 = 2k + 1$, Eq. (6.49) can be evaluated to yield

$$S_3^{2k+3}(a) = \frac{\Gamma(2k/3+1)}{3\,a^{2k/3+1}} + \frac{1}{\pi}\sum_{n=1}^{\infty}\frac{1}{n}\Bigg[3\,(-1)^k ia\frac{\partial^{2k+4}}{\partial\beta^{2k+4}}\Bigg\{\frac{z}{2\beta}\Bigg(e^{\pi i/4} \times S_{0,1/3}\Big(ze^{3\pi i/4}\Big) - e^{-\pi i/4}S_{0,1/3}\Big(ze^{-3\pi i/4}\Big)\Bigg)\Bigg\} + (-1)^k(2k+2) \times \frac{\partial^{2k+1}}{\partial\beta^{2k+1}}\Bigg\{\frac{z}{2\beta}\Bigg(e^{\pi i/4}S_{0,1/3}\Big(ze^{3\pi i/4}\Big) + e^{-\pi i/4}S_{0,1/3}\Big(ze^{-3\pi i/4}\Big)\Bigg)\Bigg\}\Bigg] \;. \tag{6.51}$$

For $r = 1$, Eq. (6.49) becomes

$$S_3(a) = \frac{\Gamma(1/3)}{3a^{1/3}} + \frac{1}{2} + \frac{2}{3}\sum_{n=1}^{\infty}\sqrt{\frac{2\pi n}{3a}}\left[e^{\pi i/4}S_{0,1/3}\left(\kappa e^{3\pi i/4}\right)\right.$$
$$\left. + \; e^{-\pi i/4}S_{0,1/3}\left(\kappa e^{-3\pi i/4}\right)\right] , \tag{6.52}$$

where $\kappa = (2/a)^{1/2}(2n\pi/3)^{3/2}$ as in Sec. 4. To avoid the direct evaluation of multiple derivatives of the Lommel functions in Eqs. (6.50) and (6.51), we can employ the relations given in Sec. 10.72 of Watson [27]. As an example, if we set $k = 0$ and utilise these relations, then we are able to obtain from Eq. (6.50)

$$S_3^2(a) = \sum_{n=1}^{\infty} ne^{-an^3} = \frac{\Gamma(2/3)}{3\,a^{2/3}} + \frac{1}{3\sqrt{3a}\,\pi}\sum_{n=1}^{\infty}\left[\sqrt{\beta}\,\left(3a/2\beta^3 - i\right)\right.$$
$$\times\; e^{3i\pi/4}S_{0,1/3}\left(ze^{-3\pi i/4}\right) - \sqrt{\beta}\left(3a/2\beta^3 + i\right)e^{i\pi/4}S_{0,1/3}\left(ze^{3\pi i/4}\right)$$
$$+\; \frac{14\sqrt{3a}}{\beta}\left(S_{-1,4/3}\left(ze^{3\pi i/4}\right) + S_{-1,4/3}\left(ze^{-3\pi i/4}\right)\right)$$
$$+\frac{40\sqrt{\beta}}{3}\left(e^{-\pi i/4}S_{-2,7/3}\left(ze^{3\pi i/4}\right) + e^{\pi i/4}S_{-2,7/3}\left(ze^{-3\pi i/4}\right)\right)$$
$$\left. +\; \frac{640\,\beta^2\,i}{27\sqrt{3a}}\left(S_{-3,10/3}\left(ze^{3\pi i/4}\right) - S_{-3,10/3}\left(ze^{-3\pi i/4}\right)\right)\right] . \tag{6.53}$$

To obtain a small a-expansion for the $p = 3$ generalised Euler-Jacobi series, we can employ the asymptotic expansion of $S_{\mu,\nu}(z)$, which is given in Sec. 10.75 of Ref. [27] and appears as No. 8.576 in Gradshteyn and Ryzhik [16]. Introducing this result into Eq. (6.52) yields

$$S_3(a) \sim \frac{\Gamma(1/3)}{3a^{1/3}} + \frac{1}{2} - \frac{1}{\pi}\sum_{n=1}^{\infty}\frac{1}{n}\sum_{m=0}^{p-1}\frac{\Gamma(m+2/3)}{\Gamma(2/3)}\;\frac{\Gamma(m+1/3)}{\Gamma(1/3)}$$
$$\times\;\frac{a^m\sin(m\pi/2)}{(2n\pi/3)^{3m}} \;+\; O(a^p) \;. \tag{6.54}$$

Utilising Gauss' multiplication formula for the gamma function (No. 8.335(2) in Gradshteyn and Ryzhik [16]) and then interchanging the sums, we find that as $p \to \infty$, Eq. (6.54) gives the Ramanujan-Berndt result and hence does not contain the exponential corrections. We already know these exist, since we have evaluated the leading term by the method of steepest descent in Sec. 4. No oscillatory exponential correction terms appear in Eq. (6.54) because the asymptotic expansions of $S_{\mu,\nu}(z)$, which appear in Ref. [27] and are quoted in Ref. [16], are not developed to the degree of accuracy to yield these subdominant contributions. We shall next look at how these terms manifest themselves.

We now evaluate $S_3(a)$ by using the inversion formula given by Eq. (2.16) but before employing Luke's hypergeometric function theory to evaluate the subdominant exponential corrections missing in Eq. (6.54), we shall express $S_3(a)$ in terms of certain special functions from which we shall determine the asymptotics. This will enable us to see how accurate the asymptotic approximation for a hypergeometric function given by Eq. (3.1) is when compared with the known asymptotics of special functions. If Eq. (3.1) were to be of limited value in determining the asymptotics, then we would have to include the already neglected subdominant terms. We shall show that the asymptotic series derived by utilising Eq. (3.1) matches that obtained via the asymptotic expansions of special functions. Hence, Eq. (3.1) produces sufficiently accurate asymptotic forms for the generalised Euler-Jacobi series.

For $p = 3$, Eq. (2.16) becomes

$$S_3(a) = \frac{\Gamma(1/3)}{3a^{1/3}} + \frac{1}{2} + \frac{2\pi}{3\sqrt{3}\,a^{1/3}} \sum_{n=1}^{\infty} \sum_{l=0}^{2} (-1)^l z^{2l/3} \times G_{1,3}^{1,1}\left(\pm z e^{i\pi/2} \,\middle|\, \begin{matrix} 0 \\ 0, -2l/3, (1-2l)/3 \end{matrix} \right)^+ , \tag{6.55}$$

where $z = (2n\pi/3)^3 a^{-1}$. To evaluate Eq. (6.55), we require

$$G_{1,3}^{1,1}\left(\pm z e^{i\pi/2} \,\middle|\, \begin{matrix} 0 \\ 0, 0, 1/3 \end{matrix} \right)^+ = \Gamma\left(\frac{2}{3}\right)^{-1} {}_0F_1\left(\frac{2}{3}; \pm iz\right)^+ , \tag{6.56}$$

$$G_{1,3}^{1,1}\left(\pm z e^{i\pi/2} \,\middle|\, \begin{matrix} 0 \\ 0, -2/3, -1/3 \end{matrix} \right)^+ = \left(\Gamma\left(\frac{4}{3}\right)\Gamma\left(\frac{5}{3}\right)\right)^{-1} \times {}_1F_2\left(1; \frac{4}{3}, \frac{5}{3}; \pm iz\right)^+ , \tag{6.57}$$

$$G_{1,3}^{1,1}\left(\pm z e^{i\pi/2} \,\middle|\, \begin{matrix} 0 \\ 0, -4/3, -1 \end{matrix} \right)^+ = -iz^{-1}\Gamma\left(\frac{4}{3}\right)^{-1} {}_0F_1\left(\frac{4}{3}; \pm iz\right)^+ , \tag{6.58}$$

which when introduced in Eq. (6.55) yields the expression given by Eq. (3.15). After some algebra beginning with the expansions of these hypergeometric functions, it can be shown that

$$G_{1,3}^{1,1}\left(\pm z e^{i\pi/2} \,\middle|\, \begin{matrix} 0 \\ 0, 0, 1/3 \end{matrix} \right)^+ = \sqrt{2}\, z^{1/6} \left[\mathrm{ber}_{-1/3}\left(2\sqrt{z}\right) - \mathrm{bei}_{-1/3}\left(2\sqrt{z}\right)\right] , \tag{6.59}$$

$$G_{1,3}^{1,1}\left(\pm z e^{i\pi/2} \,\middle|\, \begin{matrix} 0 \\ 0, -2/3, -1/3 \end{matrix} \right)^+ = \sqrt{z}\left[e^{i\pi/4}\left(\mathbf{J}_{1/3}\left(2\sqrt{z}\,e^{-i\pi/4}\right) - \mathbf{J}_{-1/3}\left(2\sqrt{z}\,e^{-i\pi/4}\right)\right) + e^{-i\pi/4}\left(\mathbf{J}_{1/3}\left(2\sqrt{z}\,e^{i\pi/4}\right) - \mathbf{J}_{-1/3}\left(2\sqrt{z}\,e^{i\pi/4}\right)\right)\right] , \tag{6.60}$$

$$G_{1,3}^{1,1}\left(\pm\, z e^{i\pi/2} \middle| \begin{matrix} 0 \\ 0, -4/3, -1 \end{matrix}\right)^{+} = \sqrt{2}\, z^{-7/6}\Big[\mathrm{bei}_{1/3}\big(2\sqrt{z}\big) - \mathrm{ber}_{1/3}\big(2\sqrt{z}\big)\Big] \,, \tag{6.61}$$

where $\mathrm{ber}_\nu(z)$ and $\mathrm{bei}_\nu(z)$ are the Kelvin functions. Introducing Eqs. (6.59) to (6.61) into Eq. (6.55) gives

$$\begin{aligned} S_3(a) = {} & \frac{\Gamma(1/3)}{3a^{1/3}} + \frac{1}{2} + \frac{2\pi}{9} \sum_{n=1}^{\infty} \sqrt{\frac{2\pi n}{a}} \Big[\sqrt{2}\,\Big(\mathrm{ber}_{-1/3}\big(2\sqrt{z}\big) \\ & - \mathrm{bei}_{-1/3}\big(2\sqrt{z}\big) + \mathrm{bei}_{1/3}\big(2\sqrt{z}\big) - \mathrm{ber}_{1/3}\big(2\sqrt{z}\big)\Big) \\ & - \; e^{i\pi/4}\,\Big(\mathbf{J}_{1/3}\big(2\sqrt{z}\,e^{-i\pi/4}\big) - \mathbf{J}_{-1/3}\big(2\sqrt{z}\,e^{-i\pi/4}\big)\Big) \\ & + \; e^{-i\pi/4}\,\Big(\mathbf{J}_{1/3}\big(2i\sqrt{z}\,e^{i\pi/4}\big) - \mathbf{J}_{-1/3}\big(2\sqrt{z}\,e^{i\pi/4}\big)\Big)\Big] \,. \end{aligned} \tag{6.62}$$

Eq. (6.62) could also have been obtained by applying the result at the bottom of p. 768 of Prudnikov et al [22] to Eq. (6.52).

From our analysis in Sec. 3 we expect the Ramanujan-Berndt result to emanate from the ${}_1F_2$ hypergeometric function in Eq. (6.57), which in turn means that the result must emanate from the Anger functions as a consequence of Eq. (6.60). Furthermore, since all the special functions in Eq. (6.62) can be expressed in terms of Bessel functions of complex argument, we can use the asymptotics for the Bessel functions to evaluate $S_3(a)$. However, we shall see shortly that some guesswork is required to select the appropriate phases for the complex variables appearing in the Bessel functions or else markedly different results ensue. This, of course, is the essence of the Stokes phenomenon.

Since we have identified the Anger functions in Eq. (6.62) as yielding the Ramanujan-Berndt result Eq. (2.2), we shall first isolate this part from the asymptotic expansion for the Anger functions before considering the problem of evaluating the subdominant exponential terms missing from Eq. (6.54). According to No. 8.583(1) of Gradshteyn and Ryzhik [16], the asymptotic expansion for the Anger functions is

$$\begin{aligned} \mathbf{J}_\nu(z) = {} & J_\nu(z) + \frac{\sin\nu\pi}{\pi z}\Bigg[\sum_{n=0}^{p-1} (-1)^n \left(\frac{2}{z}\right)^{2n} \frac{\Gamma(n+(1+\nu)/2)}{\Gamma((1+\nu)/2)} \\ & \times\; \frac{\Gamma(n+(1-\nu)/2)}{\Gamma((1-\nu)/2)} + O\left(|z|^{-2p}\right) + \frac{\nu}{z}\sum_{n=0}^{p-1} (-1)^n \left(\frac{2}{z}\right)^{2n} \\ & \times\; \frac{\Gamma(n+1+\nu/2)}{\Gamma(1+\nu/2)}\,\frac{\Gamma(n+1-\nu/2)}{\Gamma(1-\nu/2)} + \nu\, O\left(|z|^{-2p-1}\right)\Bigg] \,, \end{aligned} \tag{6.63}$$

where $|\arg z| < \pi$. If we consider the non-Bessel part of Eq. (6.63) for each of the Anger functions in $S_3(a)$, denoted here by $S_{NB}^{A}(a)$, then we get after some algebra

$$S_{NB}^{A}(a) = -\frac{i}{3}\sum_{n=1}^{\infty}\sqrt{\frac{2n}{3\pi a z}}\sum_{m=0}^{\infty}(-1)^m z^{-m}\left(e^{-3\pi i m/2} - e^{3\pi i m/2}\right) \times \frac{\Gamma(m+2/3)}{\Gamma(2/3)}\,\frac{\Gamma(m+1/3)}{\Gamma(1/3)} \ . \tag{6.64}$$

By combining these terms the double summation in Eq. (6.54) can be obtained in the limit as $p \to \infty$. Hence, the Ramanujan-Berndt result is obtained from the non-Bessel part of Eq. (6.63).

The Kelvin functions are defined by [25]

$$\mathrm{ber}_{-1/3}\big(2\sqrt{z}\big) - \mathrm{bei}_{-1/3}\big(2\sqrt{z}\big) = 2^{-1/2}\Big[e^{\pi i/4}J_{-1/3}\big(2\sqrt{z}\,e^{3\pi i/4}\big) + e^{-\pi i/4}J_{-1/3}\big(2\sqrt{z}\,e^{-3\pi i/4}\big)\Big] \ , \tag{6.65}$$

and

$$\mathrm{bei}_{1/3}\big(2\sqrt{z}\big) - \mathrm{ber}_{1/3}\big(2\sqrt{z}\big) = 2^{-1/2}\Big[e^{-3\pi i/4}J_{1/3}\big(2\sqrt{z}\,e^{3\pi i/4}\big) + e^{3\pi i/4}J_{1/3}\big(2\sqrt{z}\,e^{-3\pi i/4}\big)\Big] \ . \tag{6.66}$$

Then the square-bracketed part of Eq. (6.62) can, therefore, be written as

$$S_J = e^{\pi i/4}\Big[J_{-1/3}\big(2\sqrt{z}\,e^{3\pi i/4}\big) - J_{1/3}\big(2\sqrt{z}\,e^{3\pi i/4}\big) - J_{1/3}\big(2\sqrt{z}\,e^{-\pi i/4}\big) + J_{-1/3}\big(2\sqrt{z}\,e^{-\pi i/4}\big)\Big] + e^{3\pi i/4}\Big[J_{1/3}\big(2\sqrt{z}\,e^{-3\pi i/4}\big) - J_{-1/3}\big(2\sqrt{z}\,e^{-3\pi i/4}\big)\Big] - e^{-\pi i/4}\Big[J_{1/3}\big(2\sqrt{z}\,e^{\pi i/4}\big) - J_{-1/3}\big(2\sqrt{z}\,e^{\pi i/4}\big)\Big] . \tag{6.67}$$

If we introduce the leading order asymptotic term for the Bessel functions, e.g. see No. 8.451.1 from Gradshteyn and Ryzhik [16], then Eq. (6.67) becomes

$$S_J \sim \big(\sqrt{z}\,\pi\big)^{-1/2}\Big[e^{9\pi i/8}\sin\big(2\sqrt{z}\,e^{-3\pi i/4} - \pi/4\big) - e^{-\pi i/8} \times \sin\big(2\sqrt{z}\,e^{3\pi i/4} - \pi/4\big) - e^{3\pi i/8}\sin\big(2\sqrt{z}\,e^{-\pi i/4} - \pi/4\big) - e^{-3\pi i/8}\sin\big(2\sqrt{z}\,e^{\pi i/4} - \pi/4\big)\Big] \ . \tag{6.68}$$

After some algebra, we get

$$S_J \sim 2\big(\sqrt{z}\,\pi\big)^{-1/2}\Big(\cosh\big(\sqrt{2z}\big) + \sinh\big(\sqrt{2z}\big)\Big)\cos\big(\sqrt{2z} + 3\pi/8\big) \ , \tag{6.69}$$

which when introduced in Eq. (6.62) gives an exponentially growing result as $a \to 0$. This, therefore, cannot be correct and is a direct manifestation of not selecting the appropriate phase in the variable for the various Bessel functions appearing in Eq. (6.67). This ability of asymptotic approximations to undergo transformations as a result of a change of phase in the variable is the Stokes phenomenon. For a discussion of the Stokes phenomenon applied to Bessel functions the reader is referred to p. 202 of Watson [27].

From Sec. II. 13 of Prudnikov et al [25], or by appealing to their series representation, the Kelvin functions can also be written as

$$\mathrm{ber}_\nu\, x \pm i\, \mathrm{bei}_\nu\, x = e^{\pm \nu\pi i} J_\nu\Big(xe^{-\pi i/4}\Big) \quad , \tag{6.70}$$

so that Eq. (6.67) becomes

$$\begin{aligned} S_J = e^{\pi i/4}\Bigg[3\Big(J_{-1/3}\big(2\sqrt{z}\, e^{-\pi i/4}\big) - J_{1/3}\big(2\sqrt{z}\, e^{-\pi i/4}\big)\Big) \\ - \frac{\sqrt{3}\, i}{2}\Big(J_{-1/3}\big(2\sqrt{z}\, e^{-\pi i/4}\big) + J_{1/3}\big(2\sqrt{z}\, e^{-\pi i/4}\big)\Big)\Bigg] \\ + e^{-\pi i/4}\Bigg[3\Big(J_{-1/3}\big(2\sqrt{z}\, e^{\pi i/4}\big) - J_{1/3}\big(2\sqrt{z}\, e^{\pi i/4}\big)\Big) \\ + \frac{\sqrt{3}\, i}{2}\Big(J_{-1/3}\big(2\sqrt{z}\, e^{\pi i/4}\big) + J_{1/3}\big(2\sqrt{z}\, e^{\pi i/4}\big)\Big)\Bigg] \quad . \end{aligned} \tag{6.71}$$

Now using the asymptotic expansions for Bessel functions, we get

$$\begin{aligned} J_{-\nu}\big(2\sqrt{z}\, e^{\pm \pi i/4}\big) - J_\nu\big(2\sqrt{z}\, e^{\pm \pi i/4}\big) = -\frac{2\, e^{\mp \pi i/8}}{\pi^{3/2}\, z^{1/4}} \cos(\pi\nu) \\ \times \Bigg\{ \sum_{m=0}^{k-1} \frac{e^{\mp \pi i m/4}}{(4\sqrt{z})^m} \frac{\Gamma(m+\nu+1/2)}{\Gamma(m+1)} \Gamma(m-\nu+1/2)\, \sin(\pi\nu/2) \\ \times \sin\big(2\sqrt{z}\, e^{\pm \pi i/4} - \pi/4 - \pi m/2\big) + R_0 \Bigg\} \quad , \end{aligned} \tag{6.72}$$

and

$$\begin{aligned} J_{-\nu}\big(2\sqrt{z}\, e^{\pm \pi i/4}\big) + J_\nu\big(2\sqrt{z}\, e^{\pm \pi i/4}\big) = \frac{2\, e^{\mp \pi i/8}}{\pi^{3/2}\, z^{1/4}} \cos(\pi\nu) \\ \times \Bigg\{ \sum_{m=0}^{k-1} \frac{e^{\mp m\pi i/4}}{(4\sqrt{z})^m} \frac{\Gamma(m+\nu+1/2)}{\Gamma(m+1)} \Gamma(m-\nu+1/2)\, \cos(\pi\nu/2) \\ \times \cos\big(2\sqrt{z}\, e^{\pm \pi i/4} - \pi/4 - \pi m/2\big) + R_1 \Bigg\} \quad , \end{aligned} \tag{6.73}$$

where the remainder terms, R_0 and R_1, which are discussed in detail in Watson [27], are bounded by

$$|R_{0,1}| < \left| \frac{\Gamma(k+\nu+1/2)}{(4\sqrt{z})^k} \frac{\Gamma(k-\nu+1/2)}{\Gamma(k+1)} \right| \quad . \tag{6.74}$$

If we introduce Eqs. (6.72) and (6.73) into Eq. (6.71), then after a little algebra we get

$$S_J = \frac{3\, e^{-\sqrt{2z}}}{\pi^{3/2}\, z^{1/4}} \left\{ \sum_{m=0}^{k-1} \frac{\Gamma(m+5/6)}{(4\sqrt{z})^m} \frac{\Gamma(m+1/6)}{\Gamma(m+1)} \times \cos\left(\sqrt{2z} - \pi/8 - 3\pi m/4\right) + R_k \right\} , \tag{6.75}$$

where $|R_k| < (2\sqrt{z})^{-k} \Gamma(k+5/6)\, \Gamma(k+1/6)/\Gamma(k+1)$. Inserting Eq. (6.75) into the square-bracketed part of Eq. (6.62) yields

$$S_3(a) = \frac{\Gamma(1/3)}{3a^{1/3}} + \frac{1}{2} + \frac{1}{\sqrt{\pi}} \sum_{n=1}^{\infty} \frac{e^{-\sqrt{2z}}}{(6n\pi a)^{1/4}} \left\{ \sum_{m=0}^{k-1} \frac{\Gamma(m+5/6)}{(4\sqrt{z})^m} \times \frac{\Gamma(m+1/6)}{\Gamma(m+1)} \cos\left(\sqrt{2z} - \pi/8 - 3\pi m/4\right) + R_k \right\} , \tag{6.76}$$

where it can be seen that application of Gauss' multiplication formula for the gamma function to the $m = 0$ term of the above result produces the oscillatory decaying exponential term in Eq. (4.12). The complete asymptotic expansion for $S_3(a)$ is then obtained by adding Eq. (6.76) to Eq. (6.64).

Now we apply the hypergeometric function theory developed in Sec. 3 to evaluate $S_3(a)$. This will indicate whether the subdominant terms neglected in Eq. (3.1) should be included in our hypergeometric function approach to obtain Eq. (6.76). From our analysis in Sec. 3, we know that the L-asymptotics of the ${}_1F_2$ hypergeometric function in Eq. (6.55) yield the Ramanujan-Berndt result, which can be obtained by putting $k = 1$ in Eq. (3.27). Hence, we obtain the zeta series expression in Eq. (4.12); so all we need to do now is to evaluate the K-asymptotics.

The K-asymptotics for the ${}_0F_1$ hypergeometric function in Eq. (6.56) can be written as

$${}_0F_1\left(2/3; -e^{\pm i\pi/2} z\right)^{\pm} \sim \Gamma(2/3) \left\{ K_{0,1}\left(z e^{\pm 3i\pi/2}\right)^{\pm} \pm K_{0,1}\left(z e^{\pm i\pi/2}\right)^{\pm} \right\} . \tag{6.77}$$

In Eq. (6.77) $K_{0,1}(z)$ is given by

$$K_{0,1}(z) = \frac{1}{2\sqrt{\pi}}\, z^{-1/12}\, e^{\sqrt{2z}} \sum_{r=0}^{\infty} N_r\, z^{-r/2} , \tag{6.78}$$

since $\beta_0 = 2$ and $\gamma = -1/12$ for the above hypergeometric function. The N_r are to be determined from the recursion relation. Utilising Eq. (6.78) allows us to write Eq. (6.77) as

$${}_0F_1\left(\frac{2}{3}; \mp iz\right)^{+} \sim \frac{\Gamma(2/3)}{\sqrt{\pi}\, z^{1/12}} \left[e^{\sqrt{2z}} \sum_{r=0}^{\infty} N_r\, z^{-r/2} \cos\left(\sqrt{2z} - \frac{\pi}{24} - \frac{\pi r}{4}\right) + e^{-\sqrt{2z}} \sum_{r=0}^{\infty} N_r\, z^{-r/2} \cos\left(\sqrt{2z} - \frac{\pi}{8} - \frac{3\pi r}{4}\right) \right] , \tag{6.79}$$

where the first term is the growing term as discussed in the vicinity of Eq. (3.15). This term must cancel with the growing terms produced by the other hypergeometric functions.

By the same procedure, we can write down the K-asymptotics for the hypergeometric functions in Eqs. (6.57) and (6.58). Hence, we get

$$ {}_1F_2\left(1; \frac{4}{3}, \frac{5}{3}; \mp iz\right)^+ \sim \frac{\Gamma(4/3)}{\sqrt{\pi}} \frac{\Gamma(5/3)}{z^{3/4}} \left[e^{\sqrt{2z}} \sum_{r=0}^{\infty} N_r\, z^{-r/2} \right. $$
$$ \times \cos\left(\sqrt{2z} - \frac{3\pi}{8} - \frac{\pi r}{4}\right) + e^{-\sqrt{2z}} \sum_{r=0}^{\infty} N_r\, z^{-r/2} $$
$$ \left. \times \cos\left(\sqrt{2z} - \frac{9\pi}{8} - \frac{3\pi r}{4}\right)\right] , \tag{6.80} $$

and

$$ {}_0F_1\left(\frac{4}{3}; \mp iz\right)^- \sim \frac{\Gamma(4/3)}{\sqrt{\pi}\, z^{5/12}} \left[e^{\sqrt{2z}} \sum_{r=0}^{\infty} N_r\, z^{-r/2} \sin\left(\sqrt{2z} - \frac{5\pi}{24} - \frac{\pi r}{4}\right) \right. $$
$$ \left. - e^{-\sqrt{2z}} \sum_{r=0}^{\infty} N_r\, z^{-r/2} \sin\left(\sqrt{2z} - \frac{5\pi}{8} - \frac{\pi r}{4}\right)\right] , \tag{6.81} $$

where we know that the N_r must be the same since the growing terms must cancel. We find after introducing Eqs. (6.79) to (6.81) into Eq. (6.55) that the growing terms cancel and we are left with

$$ S_3(a) \sim \frac{\Gamma(1/3)}{3a^{1/3}} + \frac{1}{2} + 2\sqrt{\pi} \sum_{n=1}^{\infty} \frac{e^{-\sqrt{2z}}}{(6\pi na)^{1/4}} \sum_{r=0}^{\infty} N_r\, z^{-r/2} $$
$$ \times \cos\left(\sqrt{2z} - \frac{\pi}{8} - \frac{3\pi r}{4}\right) + S_3^L(a) , \tag{6.82} $$

where $S_3^L(a)$ is determined from Eq. (3.27). Eq. (6.82) is seen to be identical to Eq. (6.76) thus confirming that no subdominant terms have been neglected in obtaining Eq. (6.76) for the asymptotic expansion for $S_3(a)$.

From Eq. (3.27) we can also write $S_3^L(a)$ in the following form:

$$ S_3^L(a) = -2 \sum_{n=1}^{\infty} \int_0^{\infty} dt\, \frac{e^{-t}}{2\pi n} \sin\left(\frac{at^3}{(2\pi n)^3}\right) . \tag{6.83} $$

The integral in this equation was evaluated when we discussed $p/q = 1/3$. Utilising the result in this section we find

$$ S_3^L(a) = \frac{2i}{3\sqrt{a}} \sum_{n=1}^{\infty} \left(\frac{2\pi n}{3}\right)^{1/2} \left\{ e^{i\pi/4} S_{0,1/3}\left(\frac{2e^{i\pi/4}}{\sqrt{a}} \left(\frac{2\pi n}{3}\right)^{3/2}\right) \right. $$
$$ \left. - e^{-i\pi/4} S_{0,1/3}\left(\frac{2e^{-i\pi/4}}{\sqrt{a}} \left(\frac{2\pi n}{3}\right)^{3/2}\right)\right\} . \tag{6.84} $$

For the ${}_1F_2$ in Eqs. (6.57) and (6.80) we obtain the following recursion relation:

$$2k\, c_k = (T'(2-k) - T'(1-k))\, c_{k-1} + T'(2-k)\, c_{k-2} \ , \tag{6.85}$$

where

$$T'(t) = (t-1/6)(t-5/6)(t-3/2) \ , \tag{6.86}$$

and $2^{-k}c_k = N_k$. After a little algebra, it can be shown that Eq. (6.85) reduces to the recursion relation obtained for the ${}_0F_1$ hypergeometric functions in Eqs. (6.56) and (6.58), which is

$$c_k = T(1-k)\, c_{k-1}/2k \ , \tag{6.87}$$

where

$$T(t) = (t-1/6)(t-5/6) \ . \tag{6.88}$$

Using these results, we find that $N_0 = 1$, $N_1 = 5/144$, $N_2 = 385/41472$, etc. From Eqs. (6.87) and (6.88), we, therefore, get

$$N_k = \frac{\Gamma(k+1/6)}{2^{2k}\Gamma(k+1)} \frac{\Gamma(k+5/6)}{\Gamma(1/6)\,\Gamma(5/6)} \ , \tag{6.89}$$

which, when introduced into Eq. (6.82), yields Eq. (6.76). Hence, using the asymptotic expression given by Eq. (3.1) yields sufficiently accurate asymptotic forms for $S_{p/q}(a)$. Only terms subdominant to Eq. (6.82) have been excluded. Furthermore, we have obtained a much greater number of terms in Eq. (6.82) than was obtained by the method of steepest descent (see Eq. (4.12). This demonstrates the superiority of using the generalised Euler-Jacobi inversion formula over other asymptotic techniques. It should be noted, however, that the asymptotic terms not appearing in Eq. (4.12) can be obtained by using the method of steepest descent, albeit after considerable effort.

It should also be noted that the inner series in Eq. (6.82) is divergent once Eq. (6.89) is introduced. However, the two series obtained by splitting the cosine into two exponentials are similar to the series obtained when developing the asymptotic power series for Airy functions. See for example p. 22 of Ref. [3]. Berry has shown in Ref. [5] that the actual series obtained for Airy functions can be written as the sum of an optimally truncated series and a divergent tail with its terms approximated by a factorial divided by a power. By applying Borel summation to the latter series he shows that it can be replaced by the subdominant exponential for the Airy functions multiplied by the Stokes multiplier, now in the form of an error function. Although resummation of a divergent series in this manner provides the most direct route to the error function smoothing of the Stokes multiplier, it has been shown rigorously by Olver, Boyd and Jones in Refs. [28–30] to be valid for particular classes of integrals, where the remainder can be expressed in closed form rather than as a divergent series.

As a consequence of our study of the even cases and $p = 3$, a pattern is emerging regarding the $T(t)$ to be used in the recursion relations. Firstly, we note for p equal to k (k either even or odd), that the asymptotic form for $S_k(a)$ will consist of ${}_0F_{k-2}$ hypergeometric functions. Of course, if k is odd then there will also be one ${}_1F_{k-1}$ hypergeometric function, which we will neglect for the time being. For the ${}_0F_{k-2}$ hypergeometric functions there will be $(k-1)$ values of ω_j to be introduced in Eq. (3.9) and hence $(k-1)$ values must appear in the $T(t)$, where each value of ω_j is negative. Furthermore, it is found that the difference between neighbouring $|\omega_j|$ is $(k-1)/k$ and that the lowest value of $|\omega_j|$ is equal to $|\beta\gamma| = (k-1)|\gamma|$, where z^γ is the common power of z that is extracted from the K-asymptotics. E.g. for $p = 3$, $\gamma = -1/12$ and hence, the lowest value of $|\omega_j|$ is $1/6$. However, this value is the value of $\beta\gamma$ determined from Eq. (3.4) for the $l = 0$ hypergeometric function in Eq. (2.14) or the $l = 0$ Meijer G-function in Eq. (2.16), which has the parameters of $2/k$, $3/k$,..., $(k-1)/k$. Inserting these values into Eq. (3.4) gives $\beta\gamma = -(k-2)/2k$. Thus for $p = k$, we expect that $T(t)$ should become

$$T(t) = \left(t - \frac{(k-2)}{2k}\right)\left(t - \frac{(3k-4)}{2k}\right)\cdots\left(t - \frac{(k-2)(2k-1)}{2k}\right) \quad . \tag{6.90}$$

Furthermore, for the ${}_1F_{k-1}$ hypergeometric function arising for odd p values of the generalised Euler-Jacobi series we see that the $T(t)$ is the same as in Eq. (6.90) but is also multiplied by $(t - (k-2)(2k-1)/2k - (k-1)/k) = (t - (2k-3)/2)$. An interesting conjecture is that the $T(t)$ for successively higher order hypergeometric functions yielding the same N_r as the ${}_0F_{k-2}$ hypergeometric functions will consist of the $T(t)$ in Eq. (6.90), multiplied by factors involving the subtraction of $k/(k-1)$ from the lowest ω_j in the next lowest order hypergeometric function. For example, the next lowest order hypergeometric function to ${}_2F_k$ is ${}_1F_{k-1}$. Hence the $T(t)$ for ${}_2F_k$ might be the $T(t)$ for ${}_1F_{k-1}$ multiplied by $(t-(2k-3)/2-(k-1)/k) = (t-(2k^2-k-2)/2k)$. Thus all higher order hypergeometric functions with such $T(t)$ would possess the same N_r with their recursion relation given by Eq. (6.90). Finally, similar $T(t)$-behaviour should occur when evaluating $S^r_{p/q}(a)$.

We now consider the $p = 5$ case for the generalised Euler-Jacobi series, which can be written as

$$S_5(a) = \frac{\Gamma(1/5)}{5a^{1/5}} + \frac{1}{2} + \frac{4\pi^2}{5\sqrt{5}\,a^{1/5}} \sum_{n=1}^{\infty}\sum_{l=0}^{4} (-1)^l \, z^{2l/5}$$
$$\times \; G^{1,1}_{1,5}\left(\pm z e^{i\pi/2} \,\middle|\, \begin{matrix} 0 \\ 0, -2l/5, (1-2l)/5, (2-2l)/5, (3-2l)/5 \end{matrix} \right)^{+} \quad , \tag{6.91}$$

where $z = (2n\pi/5)^5 a^{-1}$. To evaluate Eq. (6.91), we require

$$G^{1,1}_{1,5}\left(\pm z e^{i\pi/2} \,\middle|\, \begin{matrix} 0 \\ 0,0,1/5,2/5,3/5 \end{matrix}\right)^{+} = (\Gamma(2/5)\,\Gamma(3/5))^{-1} \times \Gamma(3/5)^{-1}\, {}_0F_3\big(2/5,3/5,4/5;\pm iz\big)^{+} , \tag{6.92}$$

$$G^{1,1}_{1,5}\left(\pm z e^{i\pi/2} \,\middle|\, \begin{matrix} 0 \\ 0,-2/5,-1/5,0,1/5 \end{matrix}\right)^{+} = (\Gamma(4/5)\,\Gamma(6/5))^{-1} \times \Gamma(7/5)^{-1}\, {}_0F_3\big(4/5,6/5,7/5;\pm iz\big)^{+} , \tag{6.93}$$

$$G^{1,1}_{1,5}\left(\pm z e^{i\pi/2} \,\middle|\, \begin{matrix} 0 \\ 0,-4/5,-3/5,-2/5,-1/5 \end{matrix}\right)^{+} = (\Gamma(6/5)\,\Gamma(7/5))^{-1} \times (\Gamma(8/5)\,\Gamma(9/5))^{-1}\, {}_1F_4\big(1;6/5,7/5,8/5,9/5;\pm iz\big)^{+} , \tag{6.94}$$

$$G^{1,1}_{1,5}\left(\pm z e^{i\pi/2} \,\middle|\, \begin{matrix} 0 \\ 0,-6/5,-1,-4/5,-3/5 \end{matrix}\right)^{+} = (iz\Gamma(3/5)\,\Gamma(4/5))^{-1} \times \Gamma(6/5)^{-1}\, {}_0F_3\big(3/5,4/5,6/5;\pm iz\big)^{-} , \tag{6.95}$$

$$G^{1,1}_{1,5}\left(\pm z e^{i\pi/2} \,\middle|\, \begin{matrix} 0 \\ 0,-8/5,-7/5,-6/5-1 \end{matrix}\right)^{+} = (iz\Gamma(6/5)\,\Gamma(7/5))^{-1} \times \Gamma(8/5)^{-1}\, {}_0F_3\big(6/5,7/5,8/5;\pm iz\big)^{-} . \tag{6.96}$$

For all hypergeometric functions appearing in Eqs. (6.92) to (6.96), β_0 is equal to 4. In addition, in evaluating the K-asymptotics for the hypergeometric functions we shall choose r to equal 2 in Eq. (3.2), so that $|\arg z| \le 2\pi - \delta$. This means that we shall only need to evaluate $\Gamma^1_4(0)$ and $\Gamma^1_4(1)$ for each hypergeometric function. Furthermore, we know that the ${}_1F_4$ in Eq. (6.94) is the only contributor to the L-asymptotics of $S_5(a)$, which is evaluated by putting $k = 2$ into Eq. (3.27). Hence, we only need to consider the K-asymptotics for each hypergeometric function.

The values of γ for the hypergeometric functions in Eqs. (6.92) to (6.96) are found to be $-3/40$, $-19/40$, $-7/8$, $-11/40$ and $-27/40$ respectively. The $T(t)$ for each ${}_0F_3$ hypergeometric function is found to be

$$T(t) = (t-3/10)(t-11/10)(t-19/10)(t-27/10) , \tag{6.97}$$

whereas for the ${}_1F_4$ hypergeometric function it is given by

$$T(t) = (t - 3/10)(t - 11/10)(t - 19/10)(t - 27/10)(t - 35/10) \ . \tag{6.98}$$

In other words, the values appearing in the $T(t)$ are just the $\beta\gamma$ for each hypergeometric function. When Eq. (6.98) is introduced into Eq. (3.8), it can be shown that this recursion relation reduces to the recursion relation obtained by using Eq. (6.97), as it should, or else the coefficients N_r for the hypergeometric functions would not be identical.

In evaluating the K-asymptotics for each hypergeometric functions, we shall obtain two exponentially growing expressions and two exponentially decaying expressions, since $\cos((2m+1)\pi/2\beta_0)$ is positive for $m = 0$ and $m = 1$ and negative for $m = 2$ and $m = 3$. Because the growing terms cancel, we should be left with two decaying exponential series for the asymptotic form of $S_5(a)$. However, we shall find that the series involving $\exp(4z^{1/4}\cos(7\pi/8))$ vanishes leaving us with one exponentially decaying series in the result for $S_5(a)$. It will be seen for the $p = 7$ case that two exponentially decaying series will be obtained in the final asymptotic form.

For the hypergeometric function in Eq. (6.92), we find that $\Gamma_4^{1,0}(1)$ equals $\exp(-i\pi/5) - 2\cos(2\pi/5)$. Hence, the K-asymptotics for this hypergeometric function are

$$\begin{aligned}
{}_0F_3\left(\frac{2}{5},\frac{3}{5},\frac{4}{5};\pm iz\right)_K^+ &= \frac{\Gamma(2/5)}{(2\pi)^{3/2}}\,\frac{\Gamma(3/5)}{z^{3/40}}\,\Gamma(4/5)\sum_{r=0}^{\infty} N_r\, z^{-r/4} \\
&\times \left\{\sum_{k=0}^{1} e^{4z^{1/4}\cos\theta_5(k)} \cos\left(\theta_5(k)\left(r+\frac{3}{10}\right) - 4z^{1/4}\sin\theta_5(k)\right)\right. \\
&\quad + \left(2\cos(2\pi/5)+1\right)\sum_{k=2}^{3} e^{4z^{1/4}\cos\theta_5(k)} \\
&\quad \times \left.\cos\left(\theta_5(k)\left(r+\frac{3}{10}\right) - \frac{\pi}{5} - 4z^{1/4}\sin\theta_5(k)\right)\right\} \ .
\end{aligned} \tag{6.99}$$

For Eq. (6.93), $\Gamma_4^{1,0}(1)$ equals $-(2\cos(2\pi/5) + \exp(-4i\pi/5))$. Thus we get

$$\begin{aligned}
{}_0F_3\left(\frac{4}{5},\frac{6}{5},\frac{7}{5};\pm iz\right)_K^+ &= \frac{\Gamma(4/5)}{(2\pi)^{3/2}}\,\frac{\Gamma(6/5)}{z^{19/40}}\,\Gamma(7/5)\sum_{r=0}^{\infty} N_r\, z^{-r/4} \\
&\times \left\{\sum_{k=0}^{1} e^{4z^{1/4}\cos\theta_5(k)} \cos\left(\theta_5(k)\left(r+\frac{19}{10}\right) - 4z^{1/4}\sin\theta_5(k)\right)\right. \\
&- \sum_{k=2}^{3} e^{4z^{1/4}\cos\theta_5(k)} \left[2\cos\frac{2\pi}{5}\cos\left(\theta_5(k)\left(r+\frac{19}{10}\right) - 4z^{1/4}\sin\theta_5(k)\right)\right. \\
&+ \left.\left.\cos\left(\theta_5(k)\left(r+\frac{19}{10}\right) - \frac{4\pi}{5} - 4z^{1/4}\sin\theta_5(k)\right)\right]\right\} \ .
\end{aligned} \tag{6.100}$$

For the ${}_1F_4$ in Eq. (6.95), $\Gamma_5^{1,1}(1) = 1 + 2(\cos(\pi/5) + \cos(3\pi/5))$, and the K-asymptotics become

$$
\begin{aligned}
{}_1F_4\left(1; \frac{6}{5}, \frac{7}{5}, \frac{8}{5}, \frac{9}{5}; \pm iz\right)_K^+ &= \frac{\Gamma(6/5)}{(2\pi)^{3/2}} \frac{\Gamma(7/5)}{z^{7/8}} \Gamma(8/5)\,\Gamma(9/5) \\
\times \sum_{r=0}^{\infty} N_r\, z^{-r/4} \Bigg\{ & \sum_{k=0}^{1} e^{4z^{1/4}\cos\theta_5(k)} \cos\left(\theta_5(k)\left(r + \frac{7}{2}\right) - 4z^{1/4}\sin\theta_5(k)\right) \\
& + \Big(1 + 2\cos(\pi/5) + 2\cos(3\pi/5)\Big) \sum_{k=2}^{3} e^{4z^{1/4}\cos\theta_5(k)} \\
& \times \cos\left(\theta_5(k)\left(r + \frac{7}{2}\right) - 4z^{1/4}\sin\theta_5(k)\right)\Bigg\} \,. \qquad (6.101)
\end{aligned}
$$

For the hypergeometric function in Eq. (6.95), we find that $\Gamma_4^{1,0}(1) = \exp(-i\pi/5) - 2\cos(2\pi/5)$ and hence, we get for the K-asymptotics

$$
\begin{aligned}
i\,{}_0F_3\left(\frac{3}{5}, \frac{4}{5}, \frac{6}{5}; \pm iz\right)_K^- &= \frac{\Gamma(3/5)}{(2\pi)^{3/2}} \frac{\Gamma(4/5)}{z^{11/40}} \Gamma(6/5) \sum_{r=0}^{\infty} N_r\, z^{-r/4} \\
\times \Bigg\{ & \sum_{k=0}^{1} (-1)^k e^{4z^{1/4}\cos\theta_5(k)} \sin\left(\theta_5(k)\left(r + \frac{11}{10}\right) - 4z^{1/4}\sin\theta_5(k)\right) \\
+ \sum_{k=2}^{3} & (-1)^k e^{4z^{1/4}\cos\theta_5(k)} \left[\sin\left(\theta_5(k)\left(r + \frac{11}{10}\right) - \frac{\pi}{5} - 4z^{1/4}\sin\theta_5(k)\right)\right. \\
& \left. - 2\cos\frac{2\pi}{5} \sin\left(\theta_5(k)\left(r + \frac{11}{10}\right) - 4z^{1/4}\sin\theta_5(k)\right)\right]\Bigg\} \,. \qquad (6.102)
\end{aligned}
$$

For the final hypergeometric function given in Eq. (6.96), we find that $\Gamma_4^{1,0}(1) = 2\cos(\pi/5) - \exp(-2i\pi/5)$. Hence, the K-asymptotics for this hypergeometric function are

$$
\begin{aligned}
i\,{}_0F_3\left(\frac{6}{5}, \frac{7}{5}, \frac{8}{5}; \pm iz\right)_K^- &= \frac{\Gamma(6/5)}{(2\pi)^{3/2}} \frac{\Gamma(7/5)}{z^{27/40}} \Gamma(8/5) \sum_{r=0}^{\infty} N_r\, z^{-r/4} \\
\times \Bigg\{ & \sum_{k=0}^{1} (-1)^k e^{4z^{1/4}\cos\theta_5(k)} \sin\left(\theta_5(k)\left(r + \frac{27}{10}\right) - 4z^{1/4}\sin\theta_5(k)\right) \\
+ \sum_{k=2}^{3} & (-1)^k e^{4z^{1/4}\cos\theta_5(k)} \left[2\cos\frac{\pi}{5} \sin\left(\theta_5(k)\left(r + \frac{27}{10}\right) - 4z^{1/4}\sin\theta_5(k)\right)\right. \\
& \left. - \sin\left(\theta_5(k)\left(r + \frac{27}{10}\right) - \frac{2\pi}{5} - 4z^{1/4}\sin\theta_5(k)\right)\right]\Bigg\} \,. \qquad (6.103)
\end{aligned}
$$

In the above equations, $\theta_p(k) = (2k+1)\pi/2(p-1)$.

Now if we introduce Eqs. (6.99) to (6.103) into Eq. (6.91) and use the identity

$$2\cos(4\pi/5) + 2\cos(2\pi/5) + 1 = 0 \ , \tag{6.104}$$

then after a little algebra, we obtain the following result for $S_5(a)$:

$$S_5(a) \sim \frac{\Gamma(1/5)}{5a^{1/5}} + \frac{1}{2} + \left(\frac{2\pi}{5a}\right)^{1/8} \sum_{n=1}^{\infty} n^{-3/8} \sum_{r=0}^{\infty} N_r \, z^{-r/4} e^{4z^{1/4}\cos(5\pi/8)}$$
$$\times \ \cos\left(\frac{5\pi r}{8} + \frac{3\pi}{16} - 4z^{1/4}\sin\left(\frac{5\pi}{8}\right)\right) + 2\sum_{k=0}^{\infty} (-1)^k \, \frac{a^{2k+1}}{(2\pi)^{10k+6}}$$
$$\times \ \frac{\Gamma(10k+6)}{\Gamma(2k+2)} \, \zeta(10k+6) \ , \tag{6.105}$$

where we have included the L-asymptotics by setting k equal to 2 in Eq. (3.27). It is interesting to note that while each of the hypergeometric functions in Eq. (6.91) has exponentials of the form of $\exp(4z^{1/4}\cos((2j+1)\pi/8))$ with j equal to integers from 0 to 3, the final asymptotic form for $S_5(a)$ contains only one decaying exponential.

The recursion relation for the ${}_0F_3$ hypergeometric functions appearing in $S_5(a)$ can be obtained after introducing Eq. (6.97) in Eq. (3.7). Then we get

$$4k\, c_k = \sum_{s=1}^{3} c_{k-s} \sum_{r=0}^{4-s} \frac{(-1)^{4-s-r}}{r!\,(4-s-r)!} \, (s+r-4)(\delta_0 - 27/10)$$
$$\times \ (\delta_0 - 19/10)(\delta_0 - 11/10)(\delta_0 - 3/10) \ , \tag{6.106}$$

where we have put $\delta_0 = r - k + s$ and introduced the factor $(s+r-4)$, so that the recursion relation vanishes when $r = 4 - s$. The recursion relation for the ${}_1F_4$ hypergeometric function in Eq. (6.91) is

$$4k\, c_k = \sum_{s=1}^{4} c_{k-s} \sum_{r=0}^{4-s} \frac{(-1)^{4-s-r}}{r!\,(4-s-r)!} \, (\delta_0 - 7/2)(\delta_0 - 27/10)$$
$$\times \ (\delta_0 - 19/10)(\delta_0 - 11/10)(\delta_0 - 3/10) \ . \tag{6.107}$$

For the lower values of k, the values of c_k evaluated by using Eqs. (6.106) and (6.107) are identical. For higher values we need to show that subtracting Eq. (6.107) from Eq. (6.106) yields zero. After subtraction we get

$$\Delta = 4(1/2-k)(k-1)\, c_{k-1} + (1/2-k)\, c_{k-1} \sum_{r=0}^{3} \frac{(-1)^{3-r}}{r!\,(3-r)!}$$
$$\times \ (r-k-17/10)(r-k-9/10)(r-k-1/10)(r-k+7/10) \ , \tag{6.108}$$

and after a little algebra, it can be shown that the sum in Eq. (6.108) equals $-4(k-1)$ and hence $\Delta = 0$. Thus, the recursion relations given by Eqs. (6.106) and (6.107) are identical. A similar procedure needs to be employed to show that the recursion relations for the ${}_0F_5$ and ${}_1F_6$ hypergeometric functions in the $p = 7$ case are identical, which we leave as an exercise for the reader. Finally, we note that Eq. (6.106) reduces to

$$4k\, c_k = (9/10 - 6k + 6k^2)\, c_{k-1} + (9/5 - 49k/5 + 12k^2 - 4k^3)\, c_{k-2} + (16\,929/10\,000 - 87k/10 + 119k^2/10 - 6k^3 + k^4)\, c_{k-3} \quad , \tag{6.109}$$

and since $N_k = 4^{-k} c_k$, we find that $N_0 = 1$, $N_1 = 9/160$, $N_2 = 441/51\,200$, $N_3 = 202\,509/81\,920\,000$, etc.

We now consider the $p = 7$ case, which will exhibit the emergence of a second exponentially decaying series as we found for $p = 6$. Of all the cases considered in this work, $p = 7$ is the most complex to evaluate because of the number of distinct hypergeometric functions appearing for $S_7(a)$ and also because the evaluation of the $\Gamma_p^{1,q}$ and their complex conjugates in Eq. (3.1) is considerably more involved. From Eq. (2.16), we get

$$S_7(a) = \frac{\Gamma(1/7)}{7a^{1/7}} + \frac{1}{2} + \frac{(2\pi)^3}{7\sqrt{7}\, a^{1/7}} \sum_{n=1}^{\infty} \sum_{l=0}^{6} (-1)^l \, z^{2l/7} \times G_{1,7}^{1,1}\left(\pm z e^{i\pi/2} \middle| \begin{matrix} 0 \\ 0, -\frac{2l}{7}, \frac{(1-2l)}{7}, \frac{(2-2l)}{7}, \frac{(3-2l)}{7}, \frac{(4-2l)}{7}, \frac{(5-2l)}{7} \end{matrix} \right)^+ \quad , \tag{6.110}$$

where $z = (2n\pi/7)^7 a^{-1}$. To evaluate Eq. (6.110), we require

$$G_{1,7}^{1,1}\left(\pm z e^{i\pi/2} \middle| \begin{matrix} 0 \\ 0, 0, 1/7, 2/7, 3/7, 4/7, 5/7 \end{matrix} \right)^+ = \frac{{}_0F_5(2/7, 3/7, 4/7, 5/7, 6/7; \pm iz)^+}{\Gamma(2/7)\,\Gamma(3/7)\,\Gamma(4/7)\,\Gamma(5/7)\,\Gamma(6/7)} \quad , \tag{6.111}$$

$$G_{1,7}^{1,1}\left(\pm z e^{i\pi/2} \middle| \begin{matrix} 0 \\ 0, -2/7, -1/7, 0, 1/7, 2/7, 3/7 \end{matrix} \right)^+ = \frac{{}_0F_5(4/7, 5/7, 6/7, 8/7, 9/7; \pm iz)^+}{\Gamma(4/7)\,\Gamma(5/7)\,\Gamma(6/7)\,\Gamma(8/7)\,\Gamma(9/7)} \quad , \tag{6.112}$$

$$G_{1,7}^{1,1}\left(\pm z e^{i\pi/2} \middle| \begin{matrix} 0 \\ 0, -4/7, -3/7, -2/7, -1/7, 0, 1/7 \end{matrix} \right)^+ = \frac{{}_0F_5(6/7, 8/7, 9/7, 10/7, 11/7; \pm iz)^+}{\Gamma(6/7)\,\Gamma(8/7)\,\Gamma(9/7)\,\Gamma(10/7)\,\Gamma(11/7)} \quad , \tag{6.113}$$

$$G_{1,7}^{1,1}\left(\pm z e^{i\pi/2} \middle| \begin{matrix} 0 \\ 0, -6/7, -5/7, -4/7, -3/7, -2/7, -1/7 \end{matrix} \right)^+ = \frac{{}_1F_6(1; 8/7, 9/7, 10/7, 11/7, 12/7, 13/7; \pm iz)^+}{\Gamma(8/7)\,\Gamma(9/7)\,\Gamma(10/7)\,\Gamma(11/7)\,\Gamma(12/7)\,\Gamma(13/7)} \quad , \tag{6.114}$$

$$G^{1,1}_{1,7}\left(\pm\, ze^{i\pi/2}\,\middle|\,\begin{matrix}0\\0,-8/7,-1,-6/7,-5/7,-4/7,-3/7\end{matrix}\right)^{+}$$
$$= \frac{{}_0F_5(3/7,4/7,5/7,6/7,8/7;\pm\, iz)^{-}}{iz\Gamma(3/7)\,\Gamma(4/7)\,\Gamma(5/7)\,\Gamma(6/7)\,\Gamma(8/7)}\ , \tag{6.115}$$

$$G^{1,1}_{1,7}\left(\pm\, ze^{i\pi/2}\,\middle|\,\begin{matrix}0\\0,-10/7,-9/7,-8/7,-1,-6/7,-5/7\end{matrix}\right)^{+}$$
$$= \frac{{}_0F_5(5/7,6/7,8/7,9/7,10/7;\pm\, iz)^{-}}{iz\Gamma(5/7)\,\Gamma(6/7)\,\Gamma(8/7)\,\Gamma(9/10)\,\Gamma(10/7)}\ , \tag{6.116}$$

$$G^{1,1}_{1,7}\left(\pm\, ze^{i\pi/2}\,\middle|\,\begin{matrix}0\\0,-12/7,-11/7,-10/7,-9/7,-8/7,-1\end{matrix}\right)^{+}$$
$$= \frac{{}_0F_5(8/7,9/7,10/7,11/7,12/7;\pm\, iz)^{-}}{iz\Gamma(8/7)\,\Gamma(9/7)\,\Gamma(10/7)\,\Gamma(11/7)\,\Gamma(12/7)}\ . \tag{6.117}$$

For all the hypergeometric functions given above β_0 is equal to 6 whereas the values of γ for each of them in Eqs. (6.111) to (6.117) are $-5/84$, $-29/84$, $-53/84$, $-77/84$, $-17/84$, $-41/84$ and $-65/84$, respectively. From Eqs. (3.13) and (3.14), each ω_0 is just β_0 multiplied by the γ value for the hypergeometric functions. Since all the ω_0 form the cyclic set $\{\omega_j\}$, which is to be used in the $T(t)$ for the recursion relation, we find for the ${}_0F_5$ hypergeometric functions that

$$T(t) = (t-5/14)(t-17/14)(t-29/14)(t-41/14)(t-53/14)$$
$$\times\ (t-65/14)\ , \tag{6.118}$$

and for the ${}_1F_6$ hypergeometric function that

$$T(t) = (t-5/14)(t-17/14)(t-29/14)(t-41/14)(t-53/14)$$
$$\times\ (t-65/14)(t-77/14)\ . \tag{6.119}$$

As expected, the values of γ differ by 6/7 and we conjecture that ${}_2F_7$ hypergeometric functions with $T(t) = (t-5/14)(t-17/14)(t-29/14)(t-41/14)(t-53/14)(t-65/14)(t-77/14)(t-89/14)$ will possess the same N_r as the ${}_0F_5$ hypergeometric functions in Eq. (6.110). We leave it as an exercise to the reader to show that the recursion relation obtained by introducing the $T(t)$ into Eq. (3.7) are identical. It should also be borne in mind that the recursion relation for the ${}_2F_7$ will have a contribution from the $U(t)$ given by Eq. (3.10).

The $\Gamma^{1,0}_6(1)$ and $\Gamma^{1,1}_7(1)$ for the various hypergeometric functions in Eq. (6.91) can be determined from Eq. (6.26) while the $\Gamma^{1,0}_6(2)$ can be determined from Eq. (6.44). However, $\Gamma^{1,1}_7(2)$ for the ${}_1F_6(\alpha_1;\rho_1,\cdots,\rho_6;z)$ needs to be evaluated by using

$$\Gamma_7^{1,1}(2) = \sum_{1\le j\le k\le 6} e^{-2i\pi(\rho_j+\rho_k)} + e^{-2i\pi\alpha_1}\Gamma_6^{1,1}(1) \ , \tag{6.120}$$

which gives a value of 3 for the hypergeometric function in Eq. (6.114). It should be mentioned that the evaluation of the various gamma functions for the hypergeometric functions in the $p = 7$ case can be simplified considerably by using the identity

$$2\cos(3\pi/7) - 2\cos(2\pi/7) + 2\cos(\pi/7) - 1 = 0 \ , \tag{6.121}$$

which is analogous to Eq. (6.104) for the $p = 5$ case.

We are now in a position to evaluate the K-asymptotics for each of the hypergeometric functions appearing in Eqs. (6.111) to (6.117). Since we know that the growing exponentials, which are now of the form $\exp(6z^{1/6}\cos((2k+1)\pi/12))$ where $k = 0$, 1 and 2, cancel in the asymptotic result for $S_7(a)$, we shall not display these terms in the K-asymptotics. After some algebra, we find

$$G_{1,7}^{1,1}\left(\pm ze^{i\pi/2}\,\middle|\,\begin{matrix}0\\0,0,1/7,2/7,3/7,4/7,5/7\end{matrix}\right)_K^+ = g(z)\sum_{r=0}^{\infty} N_r\, z^{-r/6}$$
$$\times\left\{e^{6z^{1/6}\cos(7\pi/12)}\left(\left(1+2\cos\frac{\pi}{7}\right)\cos\left(\frac{7\pi r}{12}+\frac{35\pi}{168}-6z^{1/6}\sin\frac{7\pi}{12}\right)\right.\right.$$
$$\left.-\cos\left(\frac{7\pi r}{12}+\frac{83\pi}{168}-6z^{1/6}\sin\frac{7\pi}{12}\right)\right)+\left(1+2\cos\frac{2\pi}{7}\right)$$
$$\left.\times\sum_{k=4}^{5} e^{6z^{1/6}\cos\theta_7(k)}\cos\left(\theta_7(k)\left(r+\frac{5}{14}\right)-\frac{2\pi}{7}-6z^{1/6}\sin\theta_7(k)\right)\right\} , \tag{6.122}$$

$$z^{2/7}G_{1,7}^{1,1}\left(\pm ze^{i\pi/2}\,\middle|\,\begin{matrix}0\\0,-2/7-1/7,0,1/7,2/7,3/7\end{matrix}\right)_K^+ = g(z)$$
$$\times\sum_{r=0}^{\infty} N_r\, z^{-r/6}\left\{e^{6z^{1/6}\cos(7\pi/12)}\left[\cos\left(\frac{7\pi r}{12}+\frac{179\pi}{168}-6z^{1/6}\sin\frac{7\pi}{12}\right)\right.\right.$$
$$\left.+\left(1-2\cos\frac{\pi}{7}\right)\cos\left(\frac{7\pi r}{12}+\frac{203\pi}{168}-6z^{1/6}\sin\frac{7\pi}{12}\right)\right]+\left(2\cos\frac{\pi}{7}-1\right)$$
$$\left.\times\sum_{k=4}^{5} e^{6z^{1/6}\cos\theta_7(k)}\cos\left(\theta_7(k)\left(r+\frac{29}{11}\right)+\frac{\pi}{7}-6z^{1/6}\sin\theta_7(k)\right)\right\} , \tag{6.123}$$

$$z^{4/7}G_{1,7}^{1,1}\left(\pm ze^{i\pi/2}\,\middle|\,\begin{matrix}0\\0,-4/7,-3/7,-2/7,-1/7,0,1/7\end{matrix}\right)_K^+ = g(z)$$
$$\times\sum_{r=0}^{\infty} N_r\, z^{-r/6}\left\{e^{6z^{1/6}\cos(7\pi/12)}\left[2\cos\left(\frac{7\pi r}{12}+\frac{371\pi}{168}-6z^{1/6}\sin\frac{7\pi}{12}\right)\right.\right.$$

$$\times \left(\cos\frac{\pi}{7} - \cos\frac{2\pi}{7}\right) - \cos\left(\frac{7\pi r}{12} + \frac{275\pi}{168} - 6z^{1/6}\sin\frac{7\pi}{12}\right)\Bigg]$$
$$+ \sum_{k=4}^{5} e^{6z^{1/6}\cos\theta_7(k)} \Bigg[\cos\Bigg(\theta_7(k)\left(r + \frac{53}{14}\right) - 6z^{1/6}\sin\theta_7(k)\Bigg)$$
$$- 2\cos\frac{2\pi}{7}\cos\Bigg(\theta_7(k)\left(r + \frac{53}{14}\right) - \frac{\pi}{7} - 6z^{1/6}\sin\theta_7(k)\Bigg)\Bigg]\Bigg\} , \qquad (6.124)$$

$$z^{6/7} G_{1,7}^{1,1}\left(\pm ze^{i\pi/2} \,\middle|\, \begin{matrix} 0 \\ 0, -6/7, -5/7, -4/7, -3/7, -2/7, -1/7 \end{matrix}\right)_K^+$$
$$= g(z)\sum_{r=0}^{\infty} N_r z^{-r/6} \Bigg\{ 2e^{6z^{1/6}\cos(7\pi/12)}\cos\left(\frac{7\pi r}{12} + \frac{539\pi}{168} - 6z^{1/6}\sin\frac{7\pi}{12}\right)$$
$$+ 3\sum_{k=4}^{5} e^{6z^{1/6}\cos\theta_7(k)}\cos\Bigg(\theta_7(k)\left(r + \frac{77}{14}\right) - 6z^{1/6}\sin\theta_7(k)\Bigg)\Bigg\} , \qquad (6.125)$$

$$z^{8/7} G_{1,7}^{1,1}\left(\pm ze^{i\pi/2} \,\middle|\, \begin{matrix} 0 \\ 0, -8/7, -1, -6/7, -5/7, -4/7, -3/7 \end{matrix}\right)_K^+$$
$$= g(z)\sum_{r=0}^{\infty} N_r z^{-r/6} \Bigg\{ e^{6z^{1/6}\cos(7\pi/12)} \Bigg[\sin\left(\frac{7\pi r}{12} + \frac{119\pi}{168} - 6z^{1/6}\sin\frac{7\pi}{12}\right)$$
$$\times \left(2\cos\frac{\pi}{7} - 2\cos\frac{2\pi}{7}\right) + \sin\left(\frac{7\pi r}{12} + \frac{47\pi}{168} - 6z^{1/6}\sin\frac{7\pi}{12}\right)\Bigg]$$
$$- \sum_{k=4}^{5}(-1)^k e^{6z^{1/6}\cos\theta_7(k)} \Bigg[\sin\Bigg(\theta_7(k)\left(r + \frac{17}{14}\right) - 6z^{1/6}\cos\theta_7(k)\Bigg)$$
$$- 2\cos\frac{2\pi}{7}\sin\Bigg(\theta_7(k)\left(r + \frac{17}{14}\right) + \frac{\pi}{7} - 6z^{1/6}\sin\theta_7(k)\Bigg)\Bigg]\Bigg\} , \qquad (6.126)$$

$$z^{10/7} G_{1,7}^{1,1}\left(\pm ze^{i\pi/2} \,\middle|\, \begin{matrix} 0 \\ 0, -10/7, -9/7, -8/7, -1, -6/7, -5/7 \end{matrix}\right)_K^+$$
$$= g(z)\sum_{r=0}^{\infty} N_r z^{-r/6} \Bigg\{ e^{6z^{1/6}\cos(7\pi/12)} \Bigg[\sin\left(\frac{7\pi r}{12} + \frac{311\pi}{168} - 6z^{1/6}\sin\frac{7\pi}{12}\right)$$
$$- 2\left(\cos\frac{2\pi}{7} + \cos\frac{4\pi}{7}\right)\sin\left(\frac{7\pi r}{12} + \frac{287\pi}{168} - 6z^{1/6}\sin\frac{7\pi}{12}\right)\Bigg]$$
$$+ \left(1 - 2\cos\frac{\pi}{7}\right)\sum_{k=4}^{5}(-1)^k e^{6z^{1/6}\cos\theta_7(k)}$$
$$\times \sin\Bigg(\theta_7(k)\left(r + \frac{41}{14}\right) - \frac{\pi}{7} - 6z^{1/6}\sin\theta_7(k)\Bigg)\Bigg\} , \qquad (6.127)$$

and

$$
\begin{aligned}
&- z^{12/7} G_{1,7}^{1,1}\left(\pm z e^{i\pi/2} \,\middle|\, \begin{matrix} 0 \\ 0, -12/7, -11/7, -10/7, -9/7, -8/7, -1 \end{matrix} \right)_K^{+} \\
&= g(z) \sum_{r=0}^{\infty} N_r z^{-r/6} \Biggl\{ e^{6z^{1/6}\cos(7\pi/12)} \left[\sin\left(\frac{7\pi r}{12} + \frac{407\pi}{168} - 6z^{1/6} \sin\frac{7\pi}{12} \right) \right. \\
&\quad \left. - 2\left(\cos\frac{\pi}{7} + \cos\frac{3\pi}{7} \right) \sin\left(\frac{7\pi r}{12} + \frac{455\pi}{168} - 6z^{1/6} \sin\frac{7\pi}{12} \right) \right] \\
&\quad + \left(1 + 2\cos\frac{2\pi}{7} \right) \sum_{k=4}^{5} (-1)^k e^{6z^{1/6}\cos\theta_7(k)} \\
&\quad \times \sin\left(\theta_7(k)\left(r + \frac{65}{14} \right) + \frac{2\pi}{7} - 6z^{1/6} \sin\theta_7(k) \right) \Biggr\} \ , \qquad (6.128)
\end{aligned}
$$

where $g(z) = 2(2\pi)^{-5/2}/\sqrt{6}\, z^{5/84}$ and $\theta_7(k) = (2k+1)\pi/12$ by using the result immediately after Eq. (6.103). Introducing these results into Eq. (6.110) and including Eq. (3.27) with k set equal to 3, we finally obtain, after some algebra, the $p = 7$ asymptotic form for the generalised Euler-Jacobi series. This is

$$
\begin{aligned}
S_7(a) &\sim \frac{\Gamma(1/7)}{7a^{1/7}} + \frac{1}{2} + \frac{2}{\sqrt{6}} \left(\frac{2\pi}{7a} \right)^{1/12} \sum_{n=1}^{\infty} \Biggl\{ n^{-5/12} \sum_{r=0}^{\infty} N_r\, z^{-r/6} \\
&\times \left[e^{6z^{1/6}\cos(7\pi/12)} \cos\left(\frac{7\pi r}{12} + \frac{5\pi}{24} - 6z^{1/6} \sin\frac{7\pi}{12} \right) \right. \\
&\left. + e^{6z^{1/6}\cos(11\pi/12)} \cos\left(\frac{11\pi r}{12} + \frac{\pi}{24} - 6z^{1/6} \sin\frac{11\pi}{12} \right) \right] \Biggr\} \\
&+ 2\sum_{k=0}^{\infty} (-1)^{k+1} \frac{a^{2k+1}}{(2\pi)^{14k+8}} \frac{\Gamma(14k+8)}{\Gamma(2k+2)} \zeta(14k+8) \ , \qquad (6.129)
\end{aligned}
$$

where $N_0 = 1$, $N_1 = 65/1\,008$, $N_2 = 219\,307/33\,592\,320$ and $N_3 = 8\,304\,286\,951/995\,515\,121\,664$. With $c_k = 6^k N_k$ and $T(r+s-k)$ given by Eq. (6.118), the remaining N_r can be determined from

$$
6k\, c_k = \sum_{s=1}^{5} \left(\sum_{r=0}^{5-s} \frac{(-1)^{5-s-r}}{r!\,(5-s-r)!} T(r+s-k) \right) c_{k-s} \ , \qquad (6.130)
$$

which can also be written as

$$
\begin{aligned}
6k\, c_k &= (65/28 - 15k + 15k^2)\, c_{k-1} + (65/7 - 345k/7 + 60k^2 \\
&- 20k^3)\, c_{k-2} + (143\,845/5\,488 - 1\,845k/14 + 2\,505k^2/14 \\
&- 90k^3 + 15k^4)\, c_{k-3} + (67\,405/1\,372 - 666\,381k/2\,744 + 2\,490k^2/7 \\
&- 1\,535k^3/7 + 60k^4 - 6k^5)\, c_{k-4} + (348\,168\,925/7\,529\,536 \\
&- 1\,250\,385k/5\,488 + 1\,940\,577k^2/5\,488 - 3\,475k^3/14 + 2\,445k^4/28 \\
&- 15k^5 + k^6)\, c_{k-5} \ . \qquad (6.131)
\end{aligned}
$$

The reader should note the appearance of a second exponentially decaying asymptotic series subdominant to the first in $S_7(a)$, as was found for $S_6(a)$. As p/q increases, more and more subdominant exponential series appear in the complete asymptotic expansion for $S_{p/q}(a)$ when utilising the asymptotic theory presented in Sec. 3. As mentioned in the Preface, it remains to be seen whether this theory is able to provide all subdominant exponential series in a complete asymptotic expansion. We shall attempt to answer this question in the next section by assuming that Eq. (6.82) is the complete asymptotic expansion for $S_3(a)$ and hence, if numerical values of $S_3(a)$ do not agree with those obtained from Eq. (6.82), then we can only conclude that the theory does not provide all subdominant exponential series.

7. ASYMPTOTICS BEYOND ALL ORDERS

We have already discussed the meaning of 'asymptotics beyond all orders' in the Preface and so in this section we shall be concerned primarily with linking this newly emerging area of applied mathematics to the results derived in the previous sections. Although the main thrust of this work has been to develop new and worthwhile inversion formulae, albeit formally, two major questions have arisen naturally concerning the specific meaning of these results. The first of these is:

What does it mean to display all the algebraic terms in an asymptotic expansion before presenting the transcendentally small exponential terms? In essence, this concerns why we have included all the algebraic terms in our inversion formulae for odd integers of the generalised Euler-Jacobi series when it is known that these series are divergent. The conventional (Poincaré) approach considers only a finite number of leading order terms of an asymptotic expansion and then bounds the remainder (see for example Sec. 4.6 of Ref. [19]). In most cases only a few terms are evaluated, since the remainder is usually relatively small. However, when the aim is to develop an expansion that gets closer to reproducing the original function, the conventional approach is then to find the optimal number of terms before the asymptotic expansion begins to diverge, and to discard the remaining terms in the expansion by replacing them with the bound for this optimal number. This is obviously more accurate than the former approach but still cannot reproduce the original function. For example, in discussing the asymptotics of the exponential integral $Ei(z)$, Morse and Feshbach [19] show that the optimal number of terms for finding the value of $-4e^4Ei(-4)$ is either four or five and that the value they obtain from the asymptotic expansion lies somewhere between 0.875 00 and 0.781 25 with error bounds of 0.117 19 and 0.175 79 respectively. However, this entails a 5 per cent deviation from the actual value 0.825 33 and, no matter how many terms are included, one will not obtain a more accurate result. According to Dingle [3], this problem arises from an 'over-permissive prescription' which results in the discarding of the late terms of an asymptotic expansion in favour of determining an inexactly bounded remainder. As outlined in the Preface, Dingle's approach is to develop a systematic theory so that the divergent late terms in a complete asymptotic expansion can be decoded properly, for only then will it have a chance of reproducing the original function. In this section our aim is to develop an exact expression for the late terms of the algebraic series for $S_3(a)$ and in so doing, to remove the need to bound the series by an inexact remainder. In order to obtain even more accurate values for $S_3(a)$, we shall also require the subdominant or exponentially decaying series. As a consequence, we will have given meaning to including all subdominant series in the asymptotic expansions arising from our study of the generalised Euler-Jacobi inversion formula.

The second question that arises from a cursory inspection of the results presented in the previous section is based upon the assumption that the divergence of the algebraic asymptotic series can be overcome. It is:

Why present the exponential terms of an asymptotic expansion when it is obvious that for small values of a they are masked by the algebraic terms? We have already given one reason for this. Their inclusion allows us to obtain extremely accurate values for $S_3(a)$ as $a \to 0$. However, even if one does not require this accuracy, as a moves steadily away from zero, the optimal number of terms to be retained, before the algebraic series begins to diverge, decreases significantly and the exponential terms become more important numerically. Furthermore, we shall observe that the remainder for the asymptotic expansion excluding the exponential terms will not correct the deviation from the true values of $S_3(a)$. Thus, for relatively small a, but still in the domain where use of the asymptotic expansion for $S_{p/q}(a)$ is practical, the exponential terms are not at all masked by the algebraic terms and must be included if one wishes to get fairly accurate values for $S_{p/q}(a)$.

Actually, this question applies also to the even integer cases of the generalised Euler-Jacobi series for, in effect, it asks why we should include subdominant series in an asymptotic expansion. In the case of the odd integers the dominant asymptotic series was found to be algebraic, i.e. zero exponentially, while for the even integers the dominant asymptotic series was found to be exponentially decreasing. The question of masking applies as much to the even integers as to the odd ones when we include the subdominant contributions for the even integers such as in Eq. (6.45). Therefore, to summarise, subdominant terms of an asymptotic expansion need to be included if we are to have any hope of reproducing the original function and they also become important numerically as the expansion parameter moves away from its limit point.

In applying the techniques of asymptotics beyond all orders to the generalised Euler-Jacobi series, we are immediately confronted with two obstacles.

The first of these concerns the coefficients of the subdominant exponential terms in the complete asymptotic expansions. Although we have evaluated the first few coefficients of the exponential terms of $S_{p/q}(a)$ for integer values of p/q up to 7, in order to apply Dingle's techniques we need to know the functional form of all the coefficients, i.e. we require the solution of the recursion relations for each value of p/q. In this work we have managed to obtain only the functional forms of the coefficients of both the dominant and subdominant series for $p/q = 3$ and as a consequence, we shall be concerned primarily with $S_3(a)$ in this section. We shall discuss the evaluation of the recursion relations later since valuable insight will be gained from our analysis of $S_3(a)$.

The second obstacle confronting us is, whether we have all the information necessary to reproduce the original function or series, that is whether we have all subdominant exponential terms in the asymptotic expansions for each p/q value for the generalised Euler-Jacobi series. This question is, of course, linked with whether there are still more subdominant terms missing in Luke's asymptotic expansions for hypergeometric functions [8]. According to Luke, his expansions are complete and hence, one would expect no extra sub-subdominant terms, but as indicated in the Preface, this question is perhaps best resolved by applying the hyperasymptotic approach for integrals with saddle points developed by Berry and Howls to Braaksma's work [9]. In Ref. [7] Berry and Howls state that they have greatly extended the accuracy of the method of steepest descent by basing it on an exact resurgence relation. They also state that the divergence in an asymptotic series derived via the method of steepest descent is due to the existence of other saddle points , through which the path of steepest descent does not pass and that it is these saddle points that contribute the subdominant exponential terms in an asymptotic expansion. Their refinement of the method of steepest descent aims to include the subdominant contributions from the other saddle points via a unique deformation of the path of steepest descent. In his work, Braaksma states that the most difficult part is the estimation of the remainder terms for the asymptotic expansions of hypergeometric functions, but adds that they can be estimated by employing the method of steepest descent, which is sketched out by him in Sec. 10 of Ref. [9]. It may, therefore, be possible to extend his results further by using the Berry and Howls refinement of the method of steespest descent and as a consequence, resolve the question of whether there are extra sub-subdominant terms not included in the results given by Luke.

We may, however, be able to answer here the question as to whether there are further or sub-subdominant terms in this theory by assuming that we have all the necessary information in $S_3(a)$ to carry out our analysis. If our analysis turns out to be deficient, then we could attribute this to the fact that there are indeed sub-subdominant terms and hence, that Braaksma's theory needs to be extended.

In the first column of Table 2 we list various numerical values of a for which we determine values of $S_3(a)$ and $S_3(a) - \Gamma(1/3)/3a^{1/3} - 1/2$ (denoted from here on by $\mathcal{T}_3(a)$) in the accompanying columns by using Mathematica [31]. $\mathcal{T}_3(a)$ is the quantity of most interest in this section and we shall refer to it as the tail of the generalised Euler-Jacobi series. In accordance with the terminology of Sec. 3, we shall refer to the algebraic part of the tail as the L-part, denoted by an L-superscript while the exponential part will be referred to as the K-part, denoted by a K-superscript. From the table we can see immediately that the tail:

(1) is almost negligible for values of a less than 10^{-4},

(2) although increasing, it is quite small for a between 10^{-2} and 10,
and

(3) becomes sizeable when compared with $S_3(a)$ for $a \geq 10^2$, eventually approaching asymptotically a value of 0.5.

Our interest in this section is to examine the behaviour of the tail by concentrating on the interplay between the L- and K-parts. It is our aim to see if Dingle's theory of terminants can be used to remove the divergences from both the L- and K-parts, so that our final expressions will match the various values of $\mathcal{T}_3(a)$ given in Table 2. In terms of the material that we have presented so far, $\mathcal{T}_3(a)$ is given by

$$\mathcal{T}_3(a) = \mathcal{T}_3^L(a) + \mathcal{T}_3^K(a) \quad , \tag{7.1}$$

where

$$\mathcal{T}_3^L(a) = S_3^L(a) = 2\sum_{m=0}^{\infty}(-1)^{m+1}\frac{a^{2m+1}}{(2\pi)^{6m+4}}\frac{\Gamma(6m+4)}{\Gamma(2m+2)}\,\zeta(6m+4) \quad , \tag{7.2}$$

and

$$\mathcal{T}_3^K(a) = \frac{2\sqrt{\pi}}{\Gamma(1/6)\Gamma(5/6)}\sum_{n=1}^{\infty}\Bigg(\frac{e^{-(2z)^{1/2}}}{(6\pi n a)^{1/4}}\sum_{m=0}^{\infty}\frac{\Gamma(m+1/6)}{(4\sqrt{z})^m} \times\frac{\Gamma(m+5/6)}{\Gamma(m+1)}\cos\Big(\sqrt{2z}-\frac{\pi}{8}-\frac{3\pi m}{4}\Big)\Bigg) \quad . \tag{7.3}$$

As before z equals $(2n\pi/3)^3a^{-1}$.

In order to carry out our asymptotic analysis of the algebraic series we need to truncate it after $m = N - 1$ terms and denote the remainder by R_N so that

$$\mathcal{T}_3^L(a) = 2\sum_{m=0}^{N-1}(-1)^{m+1}\frac{a^{2m+1}}{(2\pi)^{6m+4}}\frac{\Gamma(6m+4)}{\Gamma(2m+2)}\,\zeta(6m+4) + R_N \quad . \tag{7.4}$$

Henceforth, we shall denote the finite series on the r.h.s. of Eq. (7.4) by $\mathcal{T}_3^L(a, N)$. The remainder can be shown to be bounded by the $m = N$ term of the series above by successively integrating Eq. (6.83) by parts. Hence,

$$|R_N| < \frac{2\,a^{2N+1}}{(2\pi)^{6N+4}}\,\frac{\Gamma(6N+4)}{\Gamma(2N+2)}\,\zeta(6N+4) \quad . \tag{7.5}$$

By setting the remainder equal to this bound, we can show that the algebraic series is asymptotic, for the ratio of the modulus of the remainder to the modulus of the Nth term in the algebraic series (denoted by u_N) is

$$\frac{|R_N|}{|u_N|} \sim \frac{3^6 a^2 N^4}{4\pi^6} \quad . \tag{7.6}$$

Eq. (7.6) shows that no matter how small a is, N will eventually become so large that the remainder, and hence the series, will diverge. For a given a, the optimal number of terms used to represent the tail occurs when the ratio Eq. (7.6) is approximately unity [19]. Thus, the optimal number of terms is roughly $(2\pi/3a)^{1/2}\pi/3$; a plot of this number versus a is presented in the figure. This graph shows that as a moves away from zero the optimal number of terms decreases significantly and that for $a > 0.1$ the number of terms used to represent the tail by the algebraic series is below five.

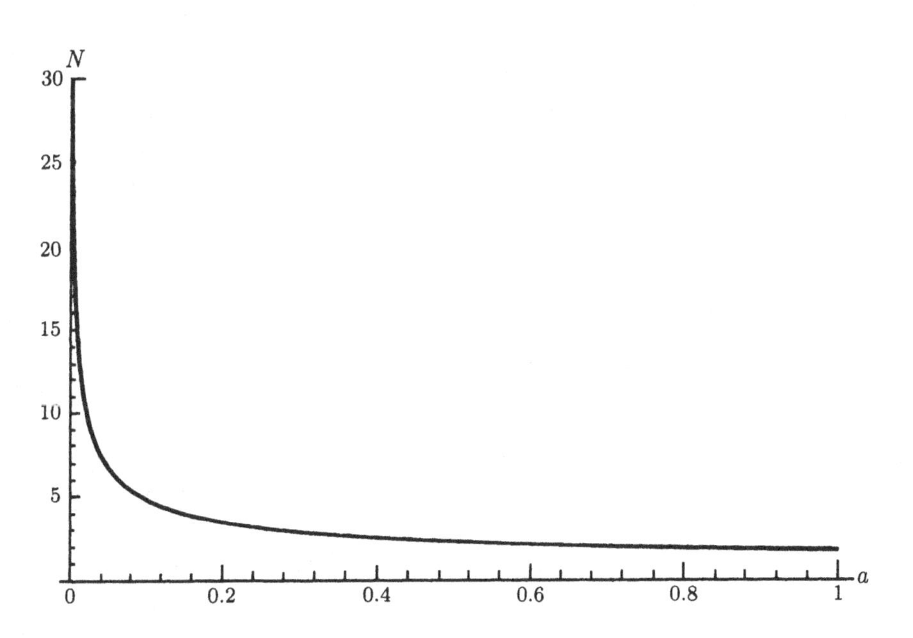

To examine how well the algebraic series approximates the tail in addition to clarifying the above discussion, we present values of $\mathcal{T}_3^L(a,N)$ as a function of N for a equal to 0.01, 0.2 and 0.4 in Tables 3, 4 and 5, respectively. In all cases the remainder decreases at first, reaches a minimum, which corresponds to the optimal number of terms, and then begins to diverge. For $a = 0.01$ the optimal number is 15 whereas for $a = 0.2$ and $a = 0.4$ it is 3 and 2 respectively. This agrees with the graph shown in the figure. On comparing this with the results in Table 2, we can see that $\mathcal{T}_3^L(0.01, 15)$ is accurate to nineteen decimal places w.r.t. $\mathcal{T}_3(a)$ while $\mathcal{T}_3^L(0.2, 2)$ is only accurate to three decimal places. Furthermore, we find that $\mathcal{T}_3^L(0.4, 1)$ is about a factor of 4 greater than the value of $\mathcal{T}_3(a)$ given in Table 2. Thus, for values of $a > 0.05$, where the deviation occurs at the fourth significant figure, terminating the algebraic series at its optimal number of terms is inadequate for yielding accurate values of the tail. This is precisely the region where the exponential terms need to be considered but in order to do so, we must remove the divergence from the algebraic series, which we know, of course, does not exist because in the previous section we were able to express the entire series as a convergent integral and then evaluate it (see Eqs. (6.83) and (6.84)).

We now consider applying Dingle's theory of terminants (see Chap. 21 of Ref. [3]) to the L- and K-parts of the tail. We begin by studying $\mathcal{T}_3^L(a)$, since we need to show that the K-part of the tail is not masked by the divergence of the late terms of $\mathcal{T}_3^L(a)$. By using the multiplication formula for the gamma function, we write Eq. (7.2) as

$$\mathcal{T}_3^L(a) = -\frac{\sqrt{3}}{2\pi^2}\sum_{n=1}^{\infty} n^{-1}\sum_{m=0}^{N_L-1}(-1)^m z^{-2m-1}\Gamma\Big(2m+\frac{4}{3}\Big) \times \Gamma\Big(2m+\frac{5}{3}\Big) - \frac{\sqrt{3}}{\pi}\sum_{n=1}^{\infty} n^{-1}\sum_{m=N_L}^{\infty}(-1)^m\left(\frac{4}{3\sqrt{3}\,z}\right)^{2m+1} \times \frac{\Gamma(m+5/6)}{\Gamma(5/6)}\,\frac{\Gamma(m+7/6)}{\Gamma(1/6)}\,\frac{\Gamma(3m+2)}{\Gamma(m+1)}\,, \tag{7.7}$$

where N_L corresponds to the optimal number of terms for a particular value of a. The second double series in Eq. (7.7) can be written more suitably as

$$\Delta\mathcal{T}_3^L(a,N_L) = \mathcal{T}_3^L(a) - \mathcal{T}_3^L(a,N_L) = -\frac{2}{3\pi^2}\sum_{n=1}^{\infty}\frac{1}{nz}\sum_{m=N_L}^{\infty}(-1)^m \times \left(\frac{16}{27z^2}\right)^m \frac{\Gamma(m+1/6)}{\Gamma(m+1)}\,\Gamma(m+5/6)\left(\frac{\Gamma(3m+3)}{3} - \frac{\Gamma(3m+2)}{2}\right)\,. \tag{7.8}$$

There are two approaches for expressing $\Delta\mathcal{T}_3^L(a)$ in terms of terminants. The first is to rewrite the quotient of gamma functions in Eq. (7.8) by using the expansion given on p. 432 of Ref. [3]. Hence, we find

$$\frac{\Gamma(m+1/6)}{\Gamma(m+1)}\,\Gamma(m+5/6)=\Gamma(m)\left(1-\frac{5}{36(m-1)}\right.$$
$$\left.+\frac{385}{2\,592\,(m-1)(m-2)}-\frac{85\,085}{279\,936\,(m-1)(m-2)(m-3)}+\;...\right)\;. \quad (7.9)$$

Thus, as m becomes very large, this quotient yields $\Gamma(m)$. This approach is essentially the one adopted by Berry in Ref. [5] and is referred to by him as the 'asymptotics of the asymptotics'. However, it is only useful if m is large, i.e. if N_L is large in $\Delta\mathcal{T}_3^L(a)$. As we have seen already, N_L is only large in the limit as $a \to 0^+$. Hence, the approach is of limited value when considering values of a greater than 0.05 and so, we must adopt a different one for the larger values of a.

In the second approach we multiply the quotient of gamma functions in Eq. (7.9) by $\Gamma(m+1/2)/\Gamma(m+1/2)$ to obtain

$$\frac{\Gamma(m+1/6)}{\Gamma(m+1)}\,\Gamma(m+5/6)=\frac{2^{2m}}{\sqrt{\pi}}\,\Gamma(m+1/2)\,B(m+1/6,m+5/6)\;. \quad (7.10)$$

Then we employ the integral representation for the beta function to express $\Delta\mathcal{T}_3^L(a)$ in terminant form. Although exact, this procedure is more complicated than the former, since it eventually leads to an increase in the dimensionality of the final terminant integrals.

In studying the asymptotics of $\mathcal{T}_3(a)$ we shall use both approaches to express the exponential part in terms of terminants. Although the first one is limited to very small values of a, it may be the only method that can be used when evaluating terminants for the exponential series of other values of p/q in the generalised Euler-Jacobi series. This is because the second approach requires an exact solution of the recursion relations that arise when studying the K-asymptotics and these may not be necessarily forthcoming. However, it is possible to evaluate asymptotic solutions of the recursion relations and hence, the first procedure can be used. Then the following problem arises:

How large must a be before the first approach becomes inaccurate?

We can gain major insight into answering this question by applying both approaches to $S_3(a)$, where we know the exact solution to the recursion relation for the exponential series, and then by comparing the results.

Let us apply the first approach to $\Delta\mathcal{T}_3^L(a,N_L)$ which now becomes

$$\Delta\mathcal{T}_3^L(a,N_L)\sim-\frac{2}{3\pi^2}\sum_{n=1}^{\infty}\frac{1}{nz}\sum_{m=N_L}^{\infty}(-1)^m\left(\frac{16}{27z^2}\right)^m\Gamma(m)$$
$$\times\left(\frac{\Gamma(3m+3)}{3}-\frac{\Gamma(3m+2)}{2}\right)\;. \quad (7.11)$$

Introducing the integral representation for the gamma function into the bracketed expression above, we find after interchanging the order of the summation and integration, that

$$\Delta\mathcal{T}_3^L(a,N_L) \sim -\frac{9a}{4\pi^5}\sum_{n=1}^{\infty}\frac{1}{n^4}\int_0^\infty dt\; e^{-t}\left(\frac{t^2}{3}-\frac{t}{2}\right) \times \sum_{m=N_L}^{\infty}(-1)^m\left(\frac{16t^3}{27z^2}\right)^m \Gamma(m) \; . \tag{7.12}$$

The sum in Eq. (7.12) is now in a form where it can be expressed in terms of terminants. Thus, using Dingle's notation, we arrive at

$$\Delta\mathcal{T}_3^L(a,N_L) \sim -\frac{9a}{4\pi^5}\sum_{n=1}^{\infty}\frac{1}{n^4}\int_0^\infty dt\; e^{-t}\left(\frac{t^2}{3}-\frac{t}{2}\right) \times \left(-\frac{16t^3}{27z^2}\right)^{N_L}\Gamma(N_L)\,\Lambda_{N_L-1}\left(\frac{27z^2}{16t^3}\right) \; , \tag{7.13}$$

where the basic terminant integral is given by

$$\Lambda_s(x) = \frac{1}{\Gamma(s+1)}\int_0^\infty dy\,\frac{y^s\, e^{-y}}{1+y/x} \; . \tag{7.14}$$

If we adopt the second approach to evaluate $\Delta\mathcal{T}_3^L(a,N_L)$, then by utilising Eq. (7.10) we find

$$\Delta\mathcal{T}_3^L(a,N_L)_E = -\frac{36a}{\sqrt{\pi}}\sum_{n=1}^{\infty}\frac{1}{(2\pi n)^4}\int_0^\infty dt\; e^{-t}\left(\frac{t^2}{3}-\frac{t}{2}\right)\int_0^1 ds\; s^{-5/6} \times (1-s)^{-1/6}\sum_{m=N_L}^{\infty}(-1)^m\left(\frac{64s(1-s)t^3}{27z^2}\right)^m\Gamma(m+1/2) \; , \tag{7.15}$$

where we have introduced the integral representation for the beta function and the subscript 'E' denotes that Eq. (7.15) is an exact result. The final sum in Eq. (7.15) is of a similar form to the m-sum found in Eq. (7.12) and hence, can also be expressed in terms of terminants. Thus, we get

$$\Delta\mathcal{T}_3^L(a,N_L)_E = -\frac{36a}{\pi^{3/2}}\sum_{n=1}^{\infty}\frac{1}{(2\pi n)^4}\int_0^\infty dt\; e^{-t}\left(\frac{t^2}{3}-\frac{t}{2}\right)\int_0^1 ds\; s^{-5/6} \times \frac{\Gamma(N_L+1/2)}{(1-s)^{1/6}}\left(\frac{64s(s-1)t^3}{27z^2}\right)^{N_L}\Lambda_{N_L-1/2}\left(\frac{27z^2}{64s(1-s)t^3}\right) \; . \tag{7.16}$$

A comparison of Eq. (7.13) with Eq. (7.16) reveals that the second approach yields more complicated higher dimensional integrals. Since Eqs. (7.13) and

(7.16) consists of sums of terminants, we shall henceforth refer to these equations as terminant sums.

We now express the above results in forms that make them amenable to numerical computation. After some manipulation, Eq. (7.13) can be written as

$$\Delta\mathcal{T}_3^L(a,N_L) \sim \left(\frac{(-\beta)^{N_L+1}\pi}{3\,a}\right)\sum_{n=1}^{\infty}\frac{1}{n^{6N_L-2}}\int_0^\infty dt\int_0^\infty dy\; e^{-(t+y)} \times \left(\frac{t^2}{3}-\frac{t}{2}\right)\left(\frac{t^{3N_L}\,y^{N_L-1}}{n^6+\beta\,t^3\,y}\right)\;, \tag{7.17}$$

while Eq. (7.16) becomes

$$\Delta\mathcal{T}_3^L(a,N_L)_E = \left(\frac{(-4\beta)^{N_L+1}}{12\,a}\right)\sqrt{\pi}\sum_{n=1}^{\infty}\frac{1}{n^{6N_L-2}}\int_0^\infty dt\int_0^\infty dy\; t^{3N_L} \times\; y^{N_L-1/2}e^{-(t+y)}\int_0^1 ds\left(\frac{t^2}{3}-\frac{t}{2}\right)\left(\frac{s^{N_L-5/6}\,(1-s)^{N_L-1/6}}{n^6+4\beta t^3\,s\,(1-s)y}\right)\quad, \tag{7.18}$$

where β equals $27a^2/4\pi^6$.

We can also apply the theory of terminants to the subdominant exponential series belonging to the tail, i.e. $\mathcal{T}_3^K(a)$, regardless of whether it converges or not. To do so, we write the cosine in Eq. (7.3) in exponential form to get

$$\mathcal{T}_3^K(a) = \sqrt{\pi}\sum_{n=1}^{\infty}\left[\frac{e^{-2\sqrt{z}\,\exp(-i\pi/4)-i\pi/8}}{(6\pi na)^{1/4}}\sum_{m=0}^{\infty}\frac{\Gamma(m+1/6)}{\Gamma(m+1)} \times\;\Gamma(m+5/6)\left(\frac{e^{-3i\pi/4}}{4\sqrt{z}}\right)^m + \frac{e^{-2\sqrt{z}\,\exp(i\pi/4)+i\pi/8}}{(6\pi na)^{1/4}}\sum_{m=0}^{\infty}\frac{\Gamma(m+1/6)}{\Gamma(m+1)} \times\;\Gamma(m+5/6)\left(\frac{e^{3i\pi/4}}{4\sqrt{z}}\right)^m\right]\;, \tag{7.19}$$

where it can be seen that the second expression on the r.h.s. of Eq. (7.19) is merely the complex conjugate of the first. Introducing N_K as the optimal number of terms to be kept in the summation over m before the series begins to diverge, we find that it can be written as

$$\mathcal{T}_3^K(a) = \mathcal{T}_3^K(a,N_K) + \Delta\mathcal{T}_3^K(a,N_K)\quad, \tag{7.20}$$

where

$$\mathcal{T}_3^K(a,N_K) = \sqrt{\pi}\sum_{n=1}^{\infty}\frac{e^{-2\sqrt{z}\,\exp(-i\pi/4)-i\pi/8}}{(6\pi na)^{1/4}}\sum_{m=0}^{N_K-1}\frac{\Gamma(m+1/6)}{\Gamma(1/6)\,\Gamma(5/6)} \times\;\frac{\Gamma(m+5/6)}{\Gamma(m+1)}\left(\frac{e^{-3i\pi/4}}{4\sqrt{z}}\right)^m + \quad \text{c.c.}\quad, \tag{7.21}$$

and

$$\Delta\mathcal{T}_3^K(a, N_K) = \sqrt{\pi} \sum_{n=1}^{\infty} \frac{e^{-2\sqrt{z}\exp(-i\pi/4)-i\pi/8}}{(6\pi na)^{1/4}} \sum_{m=N_K}^{\infty} \frac{\Gamma(m+1/6)}{\Gamma(1/6)\,\Gamma(5/6)} \times \frac{\Gamma(m+5/6)}{\Gamma(m+1)} \left(\frac{e^{-3i\pi/4}}{4\sqrt{z}}\right)^m + \text{ c.c. } , \tag{7.22}$$

where c.c. denotes the complex conjugate.

For our numerical study we need to express $\Delta\mathcal{T}_3^K(a, N_K)$ in terms of terminants by the same two methods for handling the gamma function product in Eq. (7.22) as for $\Delta\mathcal{T}_3^L(a, N_L)$. Adopting the first approach gives

$$\Delta\mathcal{T}_3^K(a, N_K) \sim \frac{\Gamma(N_K)}{2\sqrt{\pi}} \sum_{n=1}^{\infty} \frac{e^{-2\sqrt{z}\exp(-i\pi/4)-i\pi/8}}{(6\pi na)^{1/4}} \times \left(4\sqrt{z}e^{3i\pi/4}\right)^{-N_K} \Lambda_{N_K-1}\left(4\sqrt{z}\,e^{-i\pi/4}\right) + \text{ c.c. } , \tag{7.23}$$

whilst the second yields

$$\Delta\mathcal{T}_3^K(a, N_K)_E = \frac{\Gamma(N_K+1/2)}{2\pi} \sum_{n=1}^{\infty} \frac{e^{-2\sqrt{z}\exp(-i\pi/4)-i\pi/8}}{(6\pi na)^{1/4}\,(\sqrt{z}\,e^{3i\pi/4})^{N_K}} \times \int_0^{\infty} dt\; t^{N_K-5/6}\,(1-t)^{N_K-1/6}\,\Lambda_{N_K-1/2}\left(\frac{\sqrt{z}\,e^{-i\pi/4}}{t(1-t)}\right) + \text{ c.c. } , \tag{7.24}$$

where the subscript 'E' denotes that Eq. (7.24) is exact. If Eq. (7.14) is introduced into the above, then Eq. (7.23) becomes

$$\Delta\mathcal{T}_3^K(a, N_K) \sim \frac{a^{N_K/2-1/4}}{3\cdot 2^{2N_K+1/2}} \left(\frac{3}{2\pi}\right)^{3N_K/2+3/4} e^{-(3N_K+1/2)i\pi/4} \times \sum_{n=1}^{\infty} \frac{e^{-2\sqrt{z}\exp(-i\pi/4)}}{n^{3N_K/2-5/4}} \int_0^{\infty} dy\, \frac{y^{N_K-1}e^{-y}}{n^{3/2}+(e^{i\pi/4}y/4)(3/2\pi)^{3/2}\sqrt{a}} + \text{ c.c. } , \tag{7.25}$$

while Eq. (7.24) becomes

$$\Delta\mathcal{T}_3^K(a, N_K)_E = \left(\frac{3}{2\pi}\right)^{3N_K/2+1/4} \frac{a^{N_K/2-1/4}}{2\sqrt{3}\,\pi} e^{-i\pi(3N_K+1/2)/4} \times \sum_{n=1}^{\infty} \frac{e^{-2\sqrt{z}\exp(-i\pi/4)}}{n^{3N_K/2-5/4}} \int_0^1 dt\; t^{N_K-5/6}\,(1-t)^{N_K-1/6} \times \int_0^{\infty} dy\, \frac{e^{-y}\,y^{N_K-1/2}}{n^{3/2}+yt(1-t)e^{i\pi/4}(3/2\pi)^{3/2}\sqrt{a}} + \text{ c.c. } \tag{7.26}$$

We are now in a position to carry out a numerical study of the complete asymptotic expansion derived from our inversion formula for $S_3(a)$, Eq. (6.82). In this study we shall investigate the accuracy of Eq. (6.82) for a equal to

0.01, 0.2, 0.4, 1.0 and 10. Our primary aim will be to see how important the subdominant exponential series in the complete asymptotic expansion becomes as a increases from zero.

We should mention that this numerical study was carried out on an IBM 530 workstation equipped with Mathematica [31]. Initially, we used the NIntegrate routine to evaluate the terminant integrals in the various sums, but found that we could not achieve the desired accuracy expeditiously. As will be seen when discussing the hyperasymptotic stage, we will require at least 18-figure accuracy to evaluate Eq. (7.24) with $a = 0.01$. Just to achieve 7-figure accuracy using NIntegrate requires much study of the integrand so that the range of integration can be subdivided into several suitable intervals. Then, it takes an inordinate amount of CPU time for the system to yield a final result. The situation is even worse for the larger values of a. We attribute the problems encountered with using Mathematica to the substantial increase in memory capacity required to evaluate multi-dimensional integrals. As a consequence, we had to develop alternative representations for the exact terminant sums which enabled us to compute them to any desired accuracy extremely quickly. For example, to evaluate Eq. (7.24) with $a = 0.4$ to $O(10^{-12})$, i.e. to 7 significant figures, our new approach takes only 1.12 s of CPU time whereas using NIntegrate takes about 16 hr of CPU time. We present a full discussion of the numerical aspects and the new representations for the terminant sums in the next section since this is a significant advance in the subject of asymptotics beyond all orders.

In the third column of Table 6 we present values of the algebraic series terminated at its optimal value, i.e. $\mathcal{T}_3(a, N_L)$ for a equal to 0.01, 0.2, 0.4, 1.0 and 10.0 to 29 significant figures. As we shall see later, we require even more significant figures for our numerical study, which we do not present here due to limited space. Berry [20] refers to the termination of an asymptotic series at its optimal number of terms or truncation at its least term, as 'superasymptotics' because he finds for asymptotic expansions with k as the large parameter that the remainder of the dominant series is of the order of the next subdominant series rather than being of the order of $k^{-(N+1)}$ by adopting the Poincaré approach of terminating the dominant series at a fixed order N. To show this he uses the technique of resurgence whereby the late terms in the dominant series are related to the early terms in a subdominant series. Resurgence is only possible when the late terms in the dominant series are expressed as an inverse power series of what Dingle [3] describes as the 'singulant' F, which is a measure of the distance between the nearest stationary points in an integral representation for the original function. Via the Darboux theorem Dingle is able to show that the late terms of the dominant series exhibit a remarkable universality whereby the coefficients $a_r \to \Gamma(r+1-\beta)/k^r F^{r-\beta+1}$ as $r \to \infty$. Utilising this form for the coefficients, Berry is able to express the late terms of the dominant series in terms of the subdominant series and, as a consequence, obtains an error proportional to $\exp(-A|k|)$ where A is a positive constant. This hinges on the fact that the late terms in an asymptotic series take the form given above, but it can be seen from the expansions presented in the previous section that the dominant algebraic series are not of this form for large m. Hence, we cannot apply the technique of resurgence to our results. Nevertheless, we shall show that the terminant sum for the algebraic series of $S_3(a)$, given by either Eq. (7.17) or its more accurate form, Eq. (7.18), is 'superasymptotic' since the values obtained for it will be lower than the leading terms of the exponentially decaying series.

In the third column of Table 7 we give those values remaining after subtracting the values in the third column from the tail, which is essentially the difference between the actual value for $S_3(a)$ and the value obtained by truncating the algebraic series at its optimal number of terms. In light of the above we shall show that this difference is of the order of the subdominant series in the asymptotic expansion for $S_3(a)$. We can also see from this table that the superasymptotic approach works extremely well for small values of a, yielding a difference of about 7.80×10^{-20} for $a = 0.01$. However, it begins to worsen as a increases (or as N_L decreases); the difference becomes greater than 10 per cent of the actual value of $S_3(a)$ for $a \geq 0.2$. Thus, by truncating the algebraic series at its least term, vital information is lost, especially as a increases. The Poincaré specification, i.e. to truncate the algebraic series at a lower value of N, would yield even less accurate results.

From this we may ask whether the difference between the actual and the superasymptotic values for $S_3(a)$ is primarily due to the remainder of the algebraic series or whether it is primarily due to the subdominant exponential series of the complete asymptotic expansion. In Table 8 we present the $n = 1$, $n = 2$ and $n = 3$ values for both terminant sums over n in Eqs. (7.17) and (7.18) with $a = 0.01$. We denote the terms comprising the sums by $\Delta\mathcal{T}_3^L(a, N_L, n)$ for the 'asymptotics of asymptotics' approach and by $\Delta\mathcal{T}_3^L(a, N_L, n)_E$ for the exact case. The values corresponding to $\Delta\mathcal{T}_3^L(a, N_L, n)$ have been determined by using the NIntegrate routine of Mathematica and hence are only accurate to 8 figures. The values for $\Delta\mathcal{T}_3^L(a, N_L, n)_E$ have been determined by using the technique described in the next section and are considerably more accurate and faster to obtain (see Sec. 8) than their counterparts derived from the first approach. As will be seen, a very high level of accuracy for the values of $\Delta\mathcal{T}_3^L(a, N_L, n)_E$ is required so that we can carry out our numerical analysis of $S_3(a)$. For both cases we see that the $n = 1$ term is the dominant contribution to the terminant sum for the algebraic series and as a consequence, the accuracy of this term has the greatest influence on the overall accuracy of the terminant sum. In addition, we can see that both the approximate and exact approaches used to evaluate the terminant sum yield very similar results. As stated previously, this is because for very small a, N_L is sufficiently large that we can use Eq. (7.9) to zeroth order.

In Tables 9 to 12 we present the corresponding results for a equal to 0.2, 0.4, 1 and 10 respectively. For each a we have only presented the first three values in the terminant sums due to limited space. As a increases, in order to obtain sufficiently accurate values for the terminant sums one needs to evaluate far more than just three terms to obtain the overall contribution from the late terms of the algebraic series. For example, in order to achieve the required accuracy, we had to include all terms up to $n = 78$ for $a = 0.4$. In these tables we see when a is no longer very small or accordingly when N_L is no longer large, the exact approach yields different results from the 'asymptotics of the asymptotics' approach used by Berry. For the latter approach to be applicable for larger values of a, we would have to consider more terms in the expansion given by Eq. (7.9). Thus, we would end up with a series involving different terminants which would ultimately converge to the exact result given by Eq. (7.18). We shall adopt a similar approach when discussing the asymptotics of $S_4(a)$ later in this section as we are unable to produce a suitable exact solution for the corresponding recursion relation.

By comparing the results in the third column of Table 7 with the values given in Tables 8 to 12, we can see that the difference between the superasymptotic and actual values for $S_3(a)$ cannot be attributed to the remainder of the algebraic series. That is the remainder of the algebraic series truncated at its optimal number of terms is of significantly lower order than the terms in the subdominant series. For example, $\mathcal{T}_3(0.01) - \mathcal{T}_3^L(0.01, 15)$ equals 7.804×10^{-20} whereas $\Delta\mathcal{T}_3(0.01, 15, 1)_E$ equals 1.260×10^{-27}. Hence, the subdominant series in the asymptotic expansion for $S_3(a)$ must be the principal contributor to the values appearing in the third column of Table 7.

To observe this more clearly, we present numerical values of the exponentially decaying series in Eq. (6.82), i.e. $\mathcal{T}_3^K(a, N_K)$, for various values of N_K in Tables 13 to 15 with a equal to 0.01, 0.2, 0.4, 1 and 10. With the exception of the results for $\mathcal{T}_3^K(10)$, the results in these tables seem to suggest that $\mathcal{T}_3^K(a)$ is convergent: $\mathcal{T}_3^K(0.01)$ converges to $7.804\,439\,380\,266\,260\,946\,618\,30 \times 10^{-20}$ (24 figure accuracy), $\mathcal{T}_3^K(0.2)$ to $-1.694\,613\,532\,30 \times 10^{-4}$ (12 figure accuracy), $\mathcal{T}_3^K(0.4)$ to $2.412\,989\,36 \times 10^{-3}$ (9 figure accuracy), and $\mathcal{T}_3^K(1)$ to $-1.705\,77 \times 10^{-2}$. The apparent convergence is misleading because the external exponential factors are so small that they counter the early divergence of the coefficients in the m-sum. Only when m becomes extremely large will the series for $\mathcal{T}_3^K(a)$ diverge for $a < 1$. In fact, if we consider the m-sum by itself, the optimal N_K can be any value between 13 and 30 for $a = 0.4$ since the series gives values of $9.902\,460\,098 \times 10^{-1} \pm 1.0 \times 10^{-8}$ when $n = 1$. The series only begins to diverge noticeably when $N_K = 40$. For smaller a, the choice of an appropriate N_K is even more difficult. This difficulty is attributed to the oscillatory nature of the series. However, the interesting feature of both the approaches used to evaluate the terminant sums is that we can choose any value for N_K. This is particularly useful because for larger values of N_K the 'asymptotics of asymptotics' approach used by Berry becomes more accurate. We should add here that choosing a large N_K value is conditional upon having an extremely accurate method for evaluating the terminant sum by the first approach since combining the truncated exponential series at a very large value of N_K with its appropriate terminant sum is equivalent to subtracting two large and nearly equal numbers.

Aside from the fact that the major contributions to the values in the third column of Table 7 are due to the subdominant asymptotic series, we also see that the values for the terminant sums obtained by adding the values in the appropriate columns of Tables 8 to 12 are significantly lower than those produced by the subdominant series for each a. Hence, our terminant sum for the algebraic series is superasymptotic, that is by assuming only the algebraic series contributes to the asymptotic expansion of $S_3(a)$, the error one obtains by terminating or truncating the series at its optimal number of terms is less than that obtained by applying the Poincaré specification to the exponentially decaying series. However, we can still go further, thereby approaching the hyperasymptotic domain of Berry and Howls, [5–7], wherein the ultimate error is reduced from $\exp(-A|k|)$ to $\exp(-(1+2\log 2)A|k|)$. Thus, if we assume that $\exp(-A|k|)$ corresponds to the subdominant exponential in Eq. (6.82), then to reach the hyperasymptotic domain for $a = 0.01$, we would have to reduce errors to about 10^{-45} while for $a = 0.2$ and 0.4, the errors would have to be at most 10^{-10} and 10^{-8} respectively. Whether we can achieve such accuracy will ultimately depend upon whether Eq. (6.82) is the complete asymptotic expansion for $S_3(a)$ or not and upon the performance of our computer system.

In Table 16 we present the values for the remainder after both the algebraic and exponentially decaying series, optimally truncated, have been subtracted from the tail of $S_3(a)$. We denote this remainder by $\mathcal{R}_3(a, N_L, N_K)$. Thus, if Eq. (6.82) is the complete asymptotic expansion for $S_3(a)$, then the values for the remainder in Table 16 must represent the sum of the two terminant sums derived from each asymptotic series in Eq. (6.82). From here on, we shall only consider the exact values for the terminant sums since we have demonstrated previously that the first approach is not very accurate for the larger values of a.

In the third column of Table 17 we present extremely accurate values for the terminant sum of the algebraic series, $\Delta\mathcal{T}_3^L(a, N_L)_E$, which have been evaluated by using the procedure outlined in the next section. As mentioned before, using the NIntegrate routine to evaluate this quantity gives limited accuracy, e.g. for $a = 0.01$ one only gets 8 figure accuracy after 2-3 CPU hours provided appropriate cut-offs have been implemented. By comparing the results in the second column of Table 17 with those of the last column in

Table 16, we can see why we require a much higher level of accuracy than that provided by the NIntegrate routine, since our aim is to study the differences between corresponding values in the third column of Table 17.

In Table 18 we give the computed values for the exact terminant sum of the subdominant exponential series in Eq. (6.82), i.e. $\Delta\mathcal{T}_3^K(a, N_K)_E$ given by Eq. (7.18). Thus we have, once more, been able to terminate a divergent asymptotic series to yield an extremely accurate remainder and, as expected, the values for $\Delta\mathcal{T}_3^K(a, N_K)_E$ are extremely small. Most surprisingly, we see that the values given in the third column of Table 17 and the second column of Table 18 are identical. That is, there seem to be no exponentially decaying terms missing in Eq. (6.82). Thus, we conclude that although we have not conclusively proved that there are no further subdominant terms in Eq. (6.82), we have not seen any evidence of their existence for a up to 10 to a very high level of accuracy, which is consistent with Luke's statement that the asymptotic theory in Sec. 3 is complete. We would have expected to see their appearance for the larger values of a at this accuracy. Nevertheless, whether there are any further or sub-subdominant series in $S_3(a)$, or for that matter in any of the other expansions for $S_{p/q}(a)$, is a moot point. As we have stated previously, the resolution of this problem could perhaps be accomplished by developing the work of Braaksma [9] further. Nonetheless, we reiterate that the degree of accuracy in $S_3(a)$ with the existing subdominant terms is amazing.

Recapitulating the above, we have presented values for $S_3(a)$ and its tail $\mathcal{T}_3(a)$ in Table 2 for a ranging from 10^{-12} to 10^7. Here we can see that $S_3(a)$ ranges from $8.930\,3 \times 10^3$ to $1.000\,0$ (taking only the first five significant figures) whereas $\mathcal{T}_3(a)$ ranges from $-8.333\,3 \times 10^{-15}$ to $4.958\,6 \times 10^{-1}$. In Tables 3 to 5 we give the asymptotic values for $\mathcal{T}_3(a)$ by adopting the standard Poincaré approach, i.e. we neglect the subdominant series in Eq. (6.82) and approximate $S_3(a)$ by the algebraic series. We show in these tables that the algebraic series behaves as a typical asymptotic series in that after a certain number of terms it begins to diverge. In addition, as a increases, the number of optimal terms, before the series begins to diverge, reduces. The tables also provide estimates for the remainder of the algebraic series obtained by using Eq. (7.6) for fixed values of N. By adding and subtracting the error estimate for $a = 0.01$ in Table 2 to and from the optimal value for the algebraic series we should expect the actual value of $\mathcal{T}_3(a)$ to fall in the range $-8.333\,207\,107\,589\,808\,008\,820\,45 \times 10^{-5}$ to $-8.333\,207\,107\,589\,808\,008\,819\,93 \times 10^{-5}$, which according to Table 2 it does not. The cause of this can be attributed either to neglecting the subdominant asymptotic series or inadequate treatment of the remainder. This becomes more pronounced as a increases. For example, when $a = 0.2$ we see that the optimal number of terms is three and that $\mathcal{T}_3(a)$ should lie between $-1.657\,747\,682 \times 10^{-3} \pm 7.148\,19 \times 10^{-7}$ while for $a = 0.4$ we see that optimal number of terms is two and that $\mathcal{T}_3(a)$ should lie between $-3.252\,52 \times 10^{-3} \pm 3.782 \times 10^{-5}$. By observing the values for $\mathcal{T}_3(a)$ given in Table 2 we see that this is not the case. In fact, for $a = 0.4$ the upper bound of the error range is $-3.214\,70 \times 10^{-3}$, which does not compare favourably with the actual value $-8.570\,09 \times 10^{-4}$.

Although we have indicated the need to study the exponential series in the asymptotic expansion for $S_3(a)$, before doing so, we have to terminate or truncate the divergent tail of the algebraic series. We have accomplished this via two approaches; one asymptotic and the other exact; and the resulting forms for the termination of the algebraic series, which we call terminant sums, are given respectively by Eqs. (7.13) and (7.16) or in their convergent integral representation by Eqs. (7.17) and (7.18). In Table 7 we give the values remaining after the optimal number of terms for the algebraic series have been subtracted from $\mathcal{T}_3(a)$, so that we can see whether the values for the deviation between the optimal truncation of the algebraic series and the actual series can be attributed to an inadequate treatment of the late terms. In Tables 8 to 12 we present values of the some components of the terminant sums for the algebraic series obtained via both approaches. The values for the exact procedure have been obtained using the numerical procedure outlined in the next section whereas the others have been obtained by Berry's 'asymptotics of asymptotics' approach [5]. We can see for small a that the values for the components of the terminant sums compare favourably, only deviating at the third significant figure, but as a increases the deviation occurs at the first significant figure. Berry's hyperasymptotic method of iterating with optimal truncations improves the inaccuracy in using the first approach. Summing the values obtained by the exact approach for each table separately, we obtain the values for the terminant sums of the algebraic series for a equal to 0.01, 0.2, 0.4, 1 and 10. By comparing the results with their counterparts in Table 6, we can immediately observe that the values given in this table are dominated by those due to the subdominant exponential series.

Since the exponential series has the same universal behaviour for the late coefficients as the algebraic series, we are able to apply Dingle's theory of terminants to evaluate its remainder. The terminant sums for the exponential series using both the 'asymptotics of asymptotics' approach and the exact approach appear respectively in Eqs. (7.23) and (7.24) and in their convergent integral representations in Eqs. (7.25) and (7.26). We present values of the exponential series for $S_3(a)$ in Tables 13 to 15 for the various values of a considered in Table 6. With the exception of the values for a equal to 10, it is difficult to see that this series is in fact asymptotically divergent and thus difficult to choose an optimal value for truncating the series. However, unlike the work of Berry where truncation occurs at fairly large values of the asymptotic series, we do not require a large value or even the optimal value for the exact evaluation of terminant sums. That is any value will do. The advantage in choosing the optimal values for the various asymptotic series in our asymptotic expansion for $S_3(a)$ is that less numerical computation is involved. Because of the difficulty involved in finding the optimal values for each value of a, we choose arbitrary numbers of terms to truncate the exponential series, which we denote by N_K. These appear in Table 16 where we also present the values for the remainder after both the truncated algebraic and exponential series have been subtracted from $\mathcal{T}_3(a)$. Thus, if Eq. (6.82) represents the complete asymptotic expansion for $S_3(a)$, then the remainder term (denoted by $\mathcal{R}_3(a, N_L, N_K)$) must be the sum of the exact terminant sums for both series. In the second column of Table 17 we give the exact terminant sums for the algebraic series for the values of a mentioned above and then in the next column we give the remainder after subtracting these values from $\mathcal{R}_3(a, N_L, N_K)$. We then show that these values are identical to the terminant sums for the exponential series, which are presented in the second column of Table 18.

The results given in Tables 16 to 18 are nothing short of remarkable. Aside from the fact that we have tamed the divergent tails of the component asymptotic series of the asymptotic expansion for $S_3(a)$ to yield exact remainders for any value of N_L for the algebraic series and any value of N_K for the exponential series in $S_3(a)$, we have shown that our procedure can extend the range of validity of an asymptotic expansion much further away from the expansion parameter's limit point (in our case $a \to 0^+$). That is our procedure has extended the range of applicability of an asymptotic expansion to the intermediate region. Frequently in physics, many problems are solved in the limit of small and large parameter values. Thus, one ends up with a series expansion at one end and an asymptotic expansion at the other. For these situations the intermediate region can only be reached (if at all) by numerical evaluation. We have shown, however, that provided one has determined the subdominant exponential series in an asymptotic series, one can obtain a solution in a functional representation for the intermediate region. In fact, it is not necessary to determine all the subdominant terms in an asymptotic expansion, but as one moves further away from the expansion parameter's limit point, one may find after termination of the late terms of those divergent series that have been determined, that the accuracy may decline. For example, if we were not able to determine the exponential series in $S_3(a)$ as was the case when applying the Mellin transform procedure to the generalised Euler-Jacobi series (see Sec. 2), then we could still terminate the algebraic series and end up with very accurate values for $S_3(a)$ for $a < 0.01$, but these would steadily become less accurate as a approached the intermediate region. The inclusion of the subdominant exponential series in $S_3(a)$ and application of the termination procedure to the late terms of this series have allowed us to reach the intermediate region. Thus we have given meaning to the inclusion of subdominant exponential series in an asymptotic expansion. Hence, the great hope of asymptotics beyond all orders, that the largely uncharted intermediate region can be reached, has been attained. To paraphrase Dingle, asymptotics can now break free from its earlier drawbacks of vagueness and concomitant severe limitation in accuracy and range of applicability and be elevated to a discipline eliciting precise answers.

With regard to the subdominant exponential series, we have shown that no matter what the value of a is, it will always contribute to the final value of $S_{p/q}(a)$ (or for that matter any asymptotic expansion). Thus, these series should always appear in a complete asymptotic expansion, even if their contribution is small or vanishes at a particular value of a, e.g. $a = 0$ in $S_3(a)$. To reinforce this view, we can see that as we move away from the limit point $a \to 0^+$ along the real axis, the contribution of the subdominant terms increases. Furthermore, if we were to move into the complex plane, then we would see that there would be regions where the subdominant terms dominated the dominant terms, i.e. near Stokes lines. It would be incorrect to neglect the algebraic series in favour of the exponential series in these regions because to obtain the exact value for $S_3(a)$ one would require the former series even though it would yield an extremely small value. Thus we can see that the complete asymptotic expansion should always be invoked and that the values of its various component series will vary in size as one traverses the complex plane.

To apply the above theory for terminating the various exponentially decaying asymptotic series comprising the complete asymptotic expansions of the generalised Euler-Jacobi series for any value of p/q, it is apparent that we need to know the behaviour of the coefficients c_k or N_k w.r.t. k, which, in turn, means that we require the solutions to the corresponding recursion relations. This is especially important for p/q equal to an even integer since we have seen that for these cases the exponential series are the dominant part of the asymptotic expansion for $S_{p/q}(a)$. We were able to achieve extreme accuracy for $S_3(a)$ because we were able to solve the recursion relation Eq. (6.87) exactly, but what about the recursion relations for higher values of p/q, such as that given by Eq. (6.31) for $S_4(a)$? Obviously, solving equations of this type presents a formidable challenge. However, we can exploit our knowledge of the asymptotics for $S_3(a)$ to make major inroads into the solution of these more complicated recursion relations.

The first point we wish to make is that since termination of an asymptotic series is primarily concerned with its divergent tail, and hence of large values of k, we need to seek asymptotic solutions to these recursion relations. In fact, to terminate an asymptotic series we need to establish that the late terms will exhibit the fairly universal behaviour $\Gamma(k + \text{const.})/(\text{variable})^k$. For example, we saw that for $S_3(a)$ the exact solution to the recursion relation, Eq. (6.89), reduced to this form in the large k limit as evidenced by Eq. (7.9). In order to develop a methodology for handling the recursion relations for any p/q, let us begin by considering the asymptotic version of Eq. (6.31), so that we can establish that it does indeed have the given universal behaviour for large k. As a consequence, we shall determine the constant inside the gamma function and the variable, which we shall call α. Thus, for large k Eq. (6.31) reduces to

$$c_k = k\, c_{k-1} - k^2\, c_{k-2}/3 \ . \tag{7.27}$$

If we put $c_k = \Gamma(k)\alpha^k$, then we find

$$\Gamma(k)\,\alpha^k \sim \Gamma(k)\,\alpha^{k-1} - \Gamma(k)\,\alpha^{k-2}/3 \ . \tag{7.28}$$

Hence, we see that the c_k (or N_k) have the required behaviour and that the constant in the gamma function is equal to zero. In addition, the variable α is the solution of a quadratic equation and equals $\exp(\pm i\pi/6)/\sqrt{3}$. For higher values of p/q, α will become the roots of higher degree polynomials. This does not pose a problem since software packages such as Mathematica provide routines that numerically evaluate the zeros of a polynomial.

Now that we have shown that the c_k have the universal behaviour required to terminate the exponential series in the asymptotic expansion for $S_4(a)$, we put $c_k = \Gamma(k)\alpha^k M(k)$, where, in accordance with the behaviour of Eq. (7.9), $M(k)$ is given by

$$M(k) = 1 + \frac{A_1}{k-1} + \frac{A_2}{(k-1)(k-2)} + \frac{A_3}{(k-1)(k-2)(k-3)} + \frac{A_4}{(k-1)(k-2)(k-3)(k-4)} + \dots \tag{7.29}$$

The recursion relation for $S_4(a)$, therefore, becomes

$$\begin{aligned}\Gamma(k+2)\,\alpha^{k+1}M(k+1) &= \alpha^k\,\Gamma(k+2)\,M(k) + 7\alpha^k\,\Gamma(k)\,M(k)/48 \\ &- \alpha^{k-1}\,\Gamma(k+2)\,M(k-1)/3 - 7\alpha^{k-1}\,\Gamma(k)\,M(k-1)/48 \\ &- 7\alpha^{k-1}\,\Gamma(k-1)\,M(k-1)/48 \ .\end{aligned} \tag{7.30}$$

If the constant inside the gamma function for the c_k were equal to β instead of zero, then k would have to be replaced by $k+\beta$ in the above equation. By introducing Eq. (7.29) into Eq. (7.30) and Taylor-expanding the various denominators into products $(k-1)*(k-2)*\dots*(k-l)$, we obtain for the A_i in Eq. (7.29)

$$A_1 = \frac{7(1-\alpha)}{16\,(3\alpha^2-1)}, \quad A_2 = \frac{385 - 546\alpha - 287\alpha^2 + 672\alpha^3}{512\,(3\alpha^2-1)^2} \ , \tag{7.31}$$

$$\begin{aligned}A_3 = &\ \frac{7(27\,648\alpha^5 - 11\,136\alpha^4 + 31\,295\alpha^3)}{24\,576\,(3\alpha^2-1)^3} \\ &- \frac{7(12\,893\alpha^2 + 12\,387\alpha - 7\,425)}{24\,576\,(3\alpha^2-1)^3} \ ,\end{aligned} \tag{7.32}$$

$$A_4 = \frac{7(14\,598\,144\alpha^7 - 4\,893\,696\alpha^6 - 20\,218\,944\alpha^5 + 6\,073\,783\alpha^4)}{1\,572\,864\,(3\alpha^2-1)^4} + \frac{7(11\,424\,292\alpha^3 - 3\,604\,598\alpha^2 - 3\,133\,212\alpha + 1\,588\,215)}{1\,572\,864\,(3\alpha^2-1)^4}\,, \tag{7.33}$$

and so on. For the values of α given above, we find that $A_1 = -0.218\,75 \pm 0.126\,30i$, $A_2 = 0.125\,33 \pm 0.217\,07i$, $A_3 = 0.0 \pm 0.310\,53i$ and $A_4 = -0.132\,89 \pm 0.527\,60i$.

From this we have seen that for large k the coefficients c_k in the recursion relation Eq. (6.31) can be written as

$$c_k = A_0\,\alpha^k\Big[\Gamma(k) + A_1\,\Gamma(k-1) + A_2\,\Gamma(k-2) + A_3\,\Gamma(k-3) + A_4\,\Gamma(k-4) + O\Big(\Gamma(k-5)\Big)\Big] + \text{ c.c.}\,, \tag{7.34}$$

where α is taken to have positive phase and A_0 is an arbitrary complex constant that is yet to be determined. Similar results with appropriate modifications will follow for the c_k evaluated from the other recursion relations presented in this work. Since $N_k = 3^{-k}c_k$ for $S_4(a)$, Eq. (6.30) becomes

$$S_4(a) \sim \frac{\Gamma(1/4)}{4a^{1/4}} + \frac{1}{2} + \frac{2}{\sqrt{3}}\left(\frac{\pi}{2a}\right)^{1/6}\sum_{n=1}^{\infty}\sum_{k=0}^{N_K-1}\frac{N_k\,e^{-3z^{1/3}/2}}{n^{1/3}\,z^{k/3}} \times \cos\left(\frac{3\sqrt{3}}{2}\,z^{1/3} - \frac{2\pi k}{3} - \frac{\pi}{6}\right) + \frac{A_0}{\sqrt{3}}\left(\frac{\pi}{2a}\right)^{1/6}\sum_{n=1}^{\infty}\Bigg\{\frac{e^{-3z^{1/3}e^{-i\pi/3}-i\pi/6}}{n^{1/3}} \times \sum_{k=N_K}^{\infty}\left(\frac{e^{-i\pi/2}}{3\sqrt{3}\,z^{1/3}}\right)^k\Big(\Gamma(k) + A_1\Gamma(k-1) + A_2\Gamma(k-2) + A_3\Gamma(k-3) + A_4\Gamma(k-4) + \ldots\Big) + \frac{e^{-3z^{1/3}e^{i\pi/3}+i\pi/6}}{n^{1/3}}\sum_{k=N_K}^{\infty}\left(\frac{e^{5i\pi/6}}{3\sqrt{3}\,z^{1/3}}\right)^k \times \Big(\Gamma(k) + A_1\Gamma(k-1) + A_2\Gamma(k-2) + A_3\Gamma(k-3) + A_4\Gamma(k-4) + \ldots\Big)\Bigg\} + \text{ c.c.}, \tag{7.35}$$

where $N_L \gg 4$. Each of the series appearing in Eq. (7.35) can now be expressed in terms of terminants and hence we find that

$$S_4(a) \sim \frac{\Gamma(1/4)}{4a^{1/4}} + \frac{1}{2} + \frac{2}{\sqrt{3}}\left(\frac{\pi}{2a}\right)^{1/6}\sum_{n=1}^{\infty}\sum_{k=0}^{N_K-1}\frac{N_k\,e^{-3z^{1/3}/2}}{n^{1/3}\,z^{k/3}} \times \cos\left(\frac{3\sqrt{3}}{2}\,z^{1/3} - \frac{2\pi k}{3} - \frac{\pi}{6}\right) + \frac{A_0}{\sqrt{3}}\left(\frac{\pi}{2a}\right)^{1/6}\sum_{n=1}^{\infty}\Bigg\{\frac{e^{-3z^{1/3}e^{-i\pi/3}-i\pi/6}}{n^{1/3}}$$

$$\times \left(\frac{e^{-i\pi/2}}{3\sqrt{3}\,z^{1/3}}\right)^{N_K} \Big[\Gamma(N_K)\,\Lambda_{N_K-1}\big(-3\sqrt{3}\,z^{1/3}i\big) + A_1\,\Gamma(N_K-1)$$

$$\times\ \Lambda_{N_K-2}\big(-3\sqrt{3}\,z^{1/3}i\big) + A_2\,\Gamma(N_K-2)\,\Lambda_{N_K-3}\big(-3\sqrt{3}\,z^{1/3}i\big)$$

$$+\ A_3\,\Gamma(N_K-3)\,\Lambda_{N_K-4}\big(-3\sqrt{3}\,z^{1/3}i\big) + A_4\,\Gamma(N_K-4)$$

$$\times\ \Lambda_{N_K-5}\big(-3\sqrt{3}\,z^{1/3}i\big) + \,...\Big] + \frac{e^{-3z^{1/3}e^{i\pi/3}+i\pi/6}}{n^{1/3}}\left(\frac{e^{5i\pi/6}}{3\sqrt{3}\,z^{1/3}}\right)^{N_K}$$

$$\times\ \Big[\Gamma(N_K)\,\Lambda_{N_K-1}\big(3\sqrt{3}\,z^{1/3}e^{i\pi/6}\big) + A_1\,\Gamma(N_K-1)$$

$$\times\ \Lambda_{N_K-2}\big(3\sqrt{3}\,z^{1/3}e^{i\pi/6}\big) + A_2\,\Gamma(N_K-2)\,\Lambda_{N_K-3}\big(3\sqrt{3}\,z^{1/3}e^{i\pi/6}\big)$$

$$+\ A_3\,\Gamma(N_K-3)\,\Lambda_{N_K-4}\big(3\sqrt{3}\,z^{1/3}e^{i\pi/6}\big) + A_4\,\Gamma(N_K-4)$$

$$\times\ \Lambda_{N_K-5}\big(3\sqrt{3}\,z^{1/3}e^{i\pi/6}\big) + \,...\Big]\Big\} + \quad \text{c.c.}\,, \tag{7.36}$$

where c.c. means that the complex conjugate of the entire expression multiplied by A_0. The error estimates, which are obtained by considering the next order in Eq. (7.29), are

$$E \approx \frac{A_0}{\sqrt{3}}\left(\frac{\pi}{2a}\right)^{1/6}\frac{\Gamma(N_K-5)}{3^{3N_K/2}z^{N_K/3}}\sum_{n=1}^{\infty} n^{-1/3}\Big\{e^{-3z^{1/3}e^{i\pi/3}-i\pi/6-i\pi N_K/2}$$

$$\times\ \Lambda_{N_K-6}\big(-3\sqrt{3}\,z^{1/3}i\big) + e^{-3z^{1/3}e^{i\pi/3}+i\pi/6+5i\pi N_K/6}$$

$$\times\ \Lambda_{N_K-6}\big(3i\sqrt{3}\,z^{1/3}e^{i\pi/6}\big)\Big\} + \quad \text{c.c.} \tag{7.37}$$

Now all that remains is to determine the arbitrary constant A_0 in Eq. (7.37). To accomplish this we note that the values for N_k given in Table 1 can be written in terms of the general solution to the recursion relation Eq. (6.31). Thus,

$$3^{3k/2}N_k \sim \cos(k\pi/6)\ \mathrm{Re}\,(A_0)\Big(\Gamma(k) + \sum_{i=1}^{4}\mathrm{Re}\,(A_i)\,\Gamma(k-i)\Big)$$

$$+\ \sin(k\pi/6)\ \mathrm{Im}\,(A_0)\sum_{i=1}^{4}\mathrm{Im}\,(A_i)\,\Gamma(k-i)\ . \tag{7.38}$$

By using the N_{15} and N_{16} values given in the second column of Table 1, we find that $A_0 = 0.137\,830\,7 - 0.079\,578\,7i$, whereas if we use the the values for N_{19} and N_{20} we find that $A_0 = 0.137\,831\,3 - 0.079\,578\,0i$. As k increases, the more accurate A_0 becomes since the terms neglected in Eq. (7.38) become even smaller. Hence, by getting Mathematica to calculate N_k for large values of k, e.g. $k > 100$, one can obtain an extremely accurate value for A_0.

So far, we have concentrated on the asymptotics of the exponential series, which represent the dominant contributions to the complete asymptotic expansions for even integer cases of the generalised Euler-Jacobi series, but not for other values of p/q where the algebraic series dominate. Although we have succeeded in taming the divergent tail of the algebraic series for $S_3(a)$, the reader may well ask if this is an isolated case or if it is possible to obtain terminant sums for the algebraic series arising from other values of p/q. To answer this question, we turn to the algebraic series for both $S_5(a)$ and $S_7(a)$ to see if we can develop terminant sums as in the $S_3(a)$ case. If we should prove successful in developing convergent integral representations for the terminant sums, then this should convince the reader that the divergent tails of the algebraic series can be tamed, at least in principle, for all odd integer values of p/q.

From Eqs. (6.105) and (6.129) we have

$$S_5^L(a) = 2\sum_{n=1}^{\infty}\sum_{k=0}^{\infty}(-1)^k \frac{\Gamma(10k+6)}{\Gamma(2k+2)} \frac{a^{2k+1}}{(2\pi n)^{10k+6}} \; , \tag{7.39}$$

and

$$S_7^L(a) = 2\sum_{n=1}^{\infty}\sum_{k=0}^{\infty}(-1)^k \frac{\Gamma(14k+8)}{\Gamma(2k+2)} \frac{a^{2k+1}}{(2\pi n)^{14k+8}} \; . \tag{7.40}$$

By using the Gauss multiplication formula for the gamma function, we get

$$\frac{\Gamma(10k+6)}{\Gamma(2k+2)} = \frac{5^{10k+11/2}}{4\pi^2}\,\Gamma(8k+6)\,B(2k+6/5, 2k+9/5) \\ \times\; B(2k+7/5, 2k+8/5)\,B(4k+3, 4k+3) \;\; , \tag{7.41}$$

and

$$\frac{\Gamma(14+8)}{\Gamma(2k+2)} = \frac{7^{14k+15/2}}{8\pi^3}\,\Gamma(12k+9)\,B(2k+8/7, 2k+13/7) \\ \times\; B(2k+9/7, 2k+12/7)\,B(2k+10/7, 2k+11/7) \\ \times\; B(4k+3, 4k+3)\,B(4k+3, 8k+6) \;\; . \tag{7.42}$$

Introducing these results into Eqs. (7.39) and (7.40) yields

$$S_5^L(a) = 2\sum_{n=1}^{\infty}\sum_{k=0}^{N_L-1} \frac{(-1)^k a^{2k+1}}{(2\pi n)^{10k+6}} \frac{\Gamma(10k+6)}{\Gamma(2k+2)} + C_5 \sum_{n=1}^{\infty} n^{-6} \sum_{k=N_L}^{\infty} z_5^{5k/4} \\ \times\; \Gamma(k+6)\,B(k/4+6/5, k/4+9/5)\,B(k/4+7/5, k/4+8/5) \\ \times\; B(k/2+3, k/2+3) \sum_{\epsilon=\pm 1}\sum_{j=0}^{3} e^{\epsilon i\pi(2j+1)k/8} \;\; , \tag{7.43}$$

and

$$S_7^L(a) = 2\sum_{n=1}^{\infty}\sum_{k=0}^{N_L-1} \frac{(-1)^k a^{2k+1}}{(2\pi n)^{14k+8}} \frac{\Gamma(14k+8)}{\Gamma(2k+2)} + C_7 \sum_{n=1}^{\infty} n^{-8} \sum_{k=N_L}^{\infty} z_7^{7k/6}$$
$$\times\; \Gamma(k+9)\, B(k/6+8/7, k/6+13/7)\, B(k/6+9/7, k/6+12/7)$$
$$\times\; B(k/6+10/7, k/6+11/7)\, B(k/3+3, k/3+3)$$
$$\times\; B(k/3+3, 2k/3+3) \sum_{\epsilon=\pm 1}\sum_{j=0}^{5} e^{\epsilon i\pi(2j+1)k/12} \;, \qquad (7.44)$$

where $C_5 = 5^{11/2}a/4 \cdot (2\pi)^8$, $z_5 = 5a^{1/5}/2\pi n$, $C_7 = 7^{15/2}a/6 \cdot (2\pi)^{11}$ and $z_7 = 7a^{1/7}/2\pi n$.

If we use the integral representation for the beta function and Eq. (7.14), then we find that the above equations become

$$S_5^L(a) = 2\sum_{n=1}^{\infty}\sum_{k=0}^{N_L-1} (-1)^k \frac{\Gamma(10k+6)}{\Gamma(2k+2)} \frac{a^{2k+1}}{(2\pi n)^{10k+6}} + 8\, C_5 \sum_{n=1}^{\infty} n^{-6}$$
$$\times \int_0^1 \cdots \int_0^1 dt_1 \ldots dt_3 \int_0^{\infty} dy \left(\frac{y^{N_L+5}\, e^{-y}}{1+y^8\, w_5^8}\right) z_5^{5N_L/4}\, t_1^{N_L/4+1/5}$$
$$\times\; (1-t_1)^{N_L/4+4/5}\, t_2^{N_L/4+2/5}\, (1-t_2)^{N_L/2+2}\, t_3^{N_L/2+2}\, (1-t_3)^{N_L/2+2} \;, \qquad (7.45)$$

and

$$S_7^L(a) = 2\sum_{n=1}^{\infty}\sum_{k=0}^{N_L-1} (-1)^k \frac{\Gamma(14k+8)}{\Gamma(2k+2)} \frac{a^{2k+1}}{(2\pi n)^{14k+8}} + 12\, C_7 \sum_{n=1}^{\infty} n^{-8}$$
$$\times \int_0^1 \cdots \int_0^1 dt_1 \ldots dt_5 \int_0^{\infty} dy \left(\frac{y^{N_L+8}\, e^{-y}}{1+y^{12}\, w_7^{12}}\right) z_7^{7N_L/6}\, t_1^{N_L/7+1/7}$$
$$\times\; (1-t_1)^{N_L/6+6/7}\, t_2^{N_L/6+2/7}\, (1-t_2)^{N_L/6+5/7}\, t_3^{N_L/6+3/7}\, (1-t_3)^{N_L/6+4/7}$$
$$\times\; t_4^{N_L/3+2}\, (1-t_4)^{N_L/3+2}\, t_5^{N_L/3+2}\, (1-t_5)^{2N_L/3+2} \;, \qquad (7.46)$$

where $w_5 = z_5^{5/4} t_1^{1/4} (1-t_1)^{1/4} t_2^{1/4} (1-t_2)^{1/4} t_3^{1/2} (1-t_3)^{1/2}$ and $w_7 = z_7^{7/6} t_1^{1/6} (1-t_1)^{1/6} t_2^{1/6} (1-t_2)^{1/6} t_3^{1/6} (1-t_3)^{1/6} t_4^{1/3} (1-t_4)^{1/3} t_5^{1/3} (1-t_5)^{2/3}$. Thus, it is possible to construct terminant sums consisting of convergent integrals out of the late terms of the algebraic series for $S_5(a)$ and $S_7(a)$. However, this comes at the expense of increasing the dimensionality of the integrals in the resulting expressions, which makes even them less amenable to numerical computation than the integrals (both the exact and asymptotic forms) obtained for the algebraic series of $S_3(a)$. In the next section we give another derivation of the various terminant sums presented here via the Mellin-Barnes representation. These new forms greatly facilitate the numerical evaluation of terminant sums. As a consequence, we can generalise our results most readily by applying the Mellin-Barnes representation to the terminant sum of the algebraic series for arbitrary odd integer, i.e. Eq. (3.27).

8. NUMERICS FOR TERMINANT SUMS

In this section we derive exact analytical expressions for the terminant sums of the algebraic series $\mathcal{T}_3^L(a)$ given by Eq. (7.7) and the exponential series $\mathcal{T}_3^K(a)$ given by Eq. (7.20) by means of a technique which we shall call the *Mellin-Barnes regularisation of divergent series.* This technique can be used quite generally under relatively weak conditions to sum divergent series or, equivalently, to derive terminants from divergent series. As such, it represents a very powerful analytical tool in the theory of asymptotic expansions and will be expanded upon in a future publication. At first sight, the technique simply generates an alternative analytical form for the exact terminant sums, $\Delta\mathcal{T}_3^L(a, N_L)_E$ and $\Delta\mathcal{T}_3^K(a, N_K)_E$, in terms of Mellin-Barnes contour integrals. However, the replacement of the multi-dimensional integrals in Eqs. (7.18) and (7.26) by the simple contour integrals presented here greatly facilitates their numerical evaluation in terms of both speed and accuracy. As a consequence, the introduction of this transformation technique represents a major new development in the subject of asymptotics beyond all orders. As we shall see, our new forms will yield values many orders of magnitude faster and with significantly greater accuracy than those obtained by utilising standard numerical integration routines for the multi-dimensional integrals. In fact, our new procedure now competes with the direct evaluation of the original series, which is particularly important, since Berry and Howls have stated in Ref. [6] that standard numerical techniques are both more accurate and faster than asymptotics (ordinary, super- and hyper-) for large values of the expansion parameter λ (this would correspond to small values of a in our case), in solving differential equations. In addition, we reiterate that, in contrast to the expressions used in ordinary, super- and hyperasymptotics, ours are exact.

A. Mellin-Barnes regularisation

Let us begin with a general complex power series, $\sum_{k=0}^{\infty} f(k)z^k$, which may be either convergent or divergent. Such series can always be expressed as a formal contour integral via Cauchy's residue theorem as

$$\sum_{k=0}^{\infty} f(k)z^k = \frac{1}{2\pi i}\left(\int_{c-i\infty}^{c+i\infty} + \int_C\right) dt\,(-z)^t f(t)\,\Gamma(1+t)\,\Gamma(-t) \quad . \tag{8.1}$$

In Eq. (8.1) the contour C is closed to the right from $c + i\infty$ to $c - i\infty$ and includes all the poles situated along the positive real axis. The poles of $f(t)\Gamma(1+t)$ must lie to the left of the line contour through c while the poles of $\Gamma(-t)$ must lie to the right of c. Also, $|\arg(-z)| < \pi/2$.

For the cases where the series on the left diverges, we still find that the line integral $\int_{c-i\infty}^{c+i\infty}$ on the right converges provided $|\arg(-z)| < \pi/2$. Thus, all infinities due to the series have been transferred to the arc-contour C which diverges. This enables us to give a non-formal interpretation to a formal series in terms of the *finite part* of the contour integral on the r.h.s. of Eq. (8.1), i.e.

$$\sum_{k=0}^{\infty} f(k) z^k \equiv \frac{1}{2\pi i} \int_{c-i\infty}^{c+i\infty} dt\ (-z)^t f(t) \Gamma(1+t) \Gamma(-t) \ . \tag{8.2}$$

As before, the poles of $f(t)\Gamma(1+t)$ must lie to the left of the contour line through c and the poles of $\Gamma(-t)$ must lie to the right of c while $|\arg(-z)| < \pi/2$. We should mention that Eq. (8.2) could also have been obtained by a formal manipulation of each term in the series on the left via Mellin-Barnes integral-transform formulae.

Defining a *finite part* for a divergent series is somewhat analogous to the Hadamard finite part for divergent integrals in the theory of generalised functions [32]. In addition, the definition given by Eq. (8.2) yields the result one obtains by Borel summation of a divergent series. For example, consider the divergent series of

$$\sum_{k=0}^{\infty} (-z)^k \ , \quad \text{Re}\ z > 1 \ , \tag{8.3}$$

which yields $1/(1+z)$ after Borel summation. Regularising it by means of Eq. (8.2) also gives

$$\sum_{k=0}^{\infty} (-z)^k \equiv \frac{1}{2\pi i} \int_{c-i\infty}^{c+i\infty} dt\ (-z)^t\, \Gamma(1+t)\, \Gamma(-t) = \frac{1}{1+z} \ . \tag{8.4}$$

Now we come to a delicate point, which is crucial to the analysis presented in this section. It is obvious from the example above that the finite part of a divergent series is only defined, once an analytic expression, say $f(z)$, has been obtained for the right half plane, where Re $z \geq c$. Hence, a divergent series can be characterised by the expression $f(z)$. Two divergent series characterised by different analytic expressions will be different term by term since their finite parts will, in general, be different. This result is of fundamental importance and we shall now illustrate how it affects the asymptotic expansions for the generalised Euler-Jacobi series, and consequently, how it sheds new light on the dominant algebraic and subdominant exponential series comprising the complete asymptotic expansions.

Ironically, although our entire analysis began by highlighting the inadequacy of the Mellin-Barnes integral representation for the generalised Euler-Jacobi series Eq. (2.1), we now return to it, but in terms of the alternative integral representation for the tail of the series

$$S_k(a) - \frac{\Gamma(1/k)}{ka^{1/k}} - \frac{1}{2} = \frac{1}{2\pi i} \int_{1/2-i\infty}^{1/2+i\infty} dt\ a^t\, \zeta(-kt) \Gamma(-t) \ . \tag{8.5}$$

One can see that, as a result of the regularisation formula Eq. (8.2), we have simply represented the formal series for the tail by its finite part. From Eq. (2.4) the formal series for the tail is

$$S_k(a) - \frac{\Gamma(1/k)}{ka^{1/k}} - \frac{1}{2} = \sum_{l=0}^{\infty} \frac{(-a)^{l+1}}{(l+1)!} \zeta\big(-k(l+1)\big) \ . \tag{8.6}$$

We are now going to show that the formal series given above and its finite part Eq. (8.5) both contain the dominant algebraic asymptotic expansion, $\mathcal{T}_k^L(a)$, and the subdominant asymptotic exponential expansion, $\mathcal{T}_k^K(a)$, which have featured so prominently in the preceding sections. That is,

$$S_k(a) - \frac{\Gamma(1/k)}{ka^{1/k}} - \frac{1}{2} = \mathcal{T}_k^L(a) + \mathcal{T}_k^K(a) \ . \tag{8.7}$$

In fact, we shall see that $\mathcal{T}_k^L(a)$ is simply given by the even terms of the formal series, Eq. (8.6), for the tail while $\mathcal{T}_k^K(a)$ is given by the odd terms.

The dominant algebraic series for the tail of $S_3(a)$, i.e.$\mathcal{T}_3^L(a)$, has already been given by Eq. (7.7), but we can generalise this result to all positive integers by

$$\mathcal{T}_k^L(a) = \sum_{l=0}^{\infty} \frac{(-a)^{2l+1}}{(2l+1)!} \zeta\big(-k(2l+1)\big) \ . \tag{8.8}$$

Application of the Riemann reflection relation (No. 9.535(4) in Ref. [16]) to the zeta function above does not change the analytical form of the terms. The corresponding finite part for the algebraic series is obtained by the straightforward application of the regularisation formula to Eq. (8.8), which yields

$$\mathcal{T}_k^L(a) = \frac{1}{2\pi i} \int_{1/2-i\infty}^{1/2+i\infty} dt \ (-1)^{(k-1)/2} a^t \zeta(-kt) \, \Gamma(-t) \, \frac{\sin(\pi t/2)}{\sin(k\pi t/2)} \ . \tag{8.9}$$

In obtaining Eq. (8.9) we have utilised the reflection formula for the gamma function and have made a change of variable to recast the integrand. We note that when k is even, the integral in Eq. (8.9) vanishes and thus, as expected, there is no finite part for the algebraic series, $\mathcal{T}_k^L(a)$.

It should be noted that since $\zeta(-2n) = 0$, both the series for the tail, Eq. (8.6), and the algebraic series, Eq. (8.8), only differ by terms that are equal to zero. Specifically,

$$S_k(a) - \frac{\Gamma(1/k)}{ka^{1/k}} - \frac{1}{2} = \sum_{l=0}^{\infty} f(l) = a_0 + 0 + a_1 + 0 + a_2 + 0 + \dots \ ,$$

$$\mathcal{T}_k^L(a) = \sum_{l=0}^{\infty} g(l) = a_0 + a_1 + a_2 + \dots \ . \tag{8.10}$$

Yet we know that the two contributions yield different asymptotic results. How can this be? The answer lies in our observation that the analytic expression for all the terms comprising the tail of the generalised Euler-Jacobi series, $f(l)$, and the corresponding analytic expression for all the terms comprising the algebraic series, $g(l)$, are, in fact, different functions. That is, since both formal series are different, they will have different finite parts, which is easily confirmed by a close inspection of Eqs. (8.5) and (8.9).

Examples of series that differ only by an infinite number of zeros from each other, but have different finite parts, have been around for a long time. A classic example is provided by the two series

$$\begin{aligned} 1-1+1-1+1-1+1-1+\ldots &= \frac{1}{2} \ , \\ 1+0-1+1+0-1+1+0-\ldots &= \frac{2}{3} \ . \end{aligned} \tag{8.11}$$

This apparent paradox was first discovered by Callet and resolved by Lagrange using Euler's method or principle of summation for this class of divergent series. The reader should consult Ch. 1 of Hardy's classic tome [13] about this matter. In future work we shall show via another method that periodic series such as those above converge to a limit, but not according to the standard definition of convergence where the inclusion of an infinite number of zeros in a series has no effect on the limit.

Because the finite part for the tail of the generalised Euler-Jacobi series and the finite part for the algebraic series are different, we now require a subdominant or exponential series, in addition to the algebraic series, to provide the complete asymptotic expansion. We have just seen from Eqs. (8.6) and (8.8) that the algebraic series is given formally by the even terms of the series for the tail of the generalised Euler-Jacobi series. Thus, the missing subdominant part must be formally given by the odd terms of the tail, which are

$$\mathcal{T}_k^K(a) = \sum_{l=0}^{\infty} \frac{(-a)^{2l+2}}{(2l+2)!}\, \zeta\Big(-k(2l+2)\Big) \ . \tag{8.12}$$

On a formal level, all terms in this series are equal to zero. In light of the above discussion, the series yields a non-zero finite part.

By employing this regularisation procedure, we are now able to give the finite part of the exponential series, $\mathcal{T}_k^K(a)$, in terms of a Mellin-Barnes integral. This is obtained by subtracting Eq. (8.9) from Eq. (8.5), which yields

$$\begin{aligned} \mathcal{T}_k^K(a) = &\frac{1}{2\pi i} \int_{1/2-i\infty}^{1/2+i\infty} dt\, a^t\, \zeta(-kt)\, \Gamma(-t) \\ &\times \left[1 - \frac{(-1)^{(k-1)/2} \sin(\pi t/2)}{\sin(k\pi t/2)}\right] \ . \end{aligned} \tag{8.13}$$

We note that Eq. (8.13) constitutes the entire subdominant contribution to the tail of the generalised Euler-Jacobi series. Even though it has not been established conclusively that the subdominant exponential asymptotic expansions, $\mathcal{T}_k^K(a)$, obtained from the asymptotics given in Ref. [8] and presented in Sec. 3, are complete, we have not found any numerical discrepancy between the two forms, as will be demonstrated soon. This, therefore, allows us to use the same symbol, $\mathcal{T}_k^K(a)$, for both the 'Luke' and the 'complete' subdominant expansion. That is, our results are consistent with Luke's statement that his asymptotic expansions for hypergeometric functions are complete.

Eq. (8.9) for the algebraic series, $\mathcal{T}_k^L(a)$, and Eq. (8.13) for the exponential series, $\mathcal{T}_k^K(a)$, are remarkable, exact, analytical results for the finite parts of the expansions of the tail of the generalised Euler-Jacobi series. This means that the regularisation formula, Eq. (8.2), is a very general and powerful tool to derive exact analytical expressions for the terminant sums of divergent series. We believe that the integral in the regularisation formula converges whenever it is possible to obtain closed approximate forms for the finite parts of terminant sums. As a matter of fact, whenever we talk about a divergent series from here on, we shall actually be referring to its 'finite part'.

It should also be stated that the power of our Mellin-Barnes representation goes beyond producing exact analytical expressions. As it turns out, the contour integrals in Eqs. (8.9) and (8.13) not only converge, but converge so fast that they actually become the preferred choice for any numerical evaluation of terminants! We now describe this feature for $p/q = 3$ in the generalised Euler-Jacobi series which we studied extensively in the previous section.

B. Terminant sums for $S_3(a)$

For the reader to make better contact with the expressions used in the previous section, we shall not actually employ Eqs. (8.9) and (8.13) directly, but rather derive equivalent forms by applying the regularisation formula, Eq. (8.2), to the formal series for $\mathcal{T}_3^L(a)$ and $\mathcal{T}_3^K(a)$ given in that section. Of course, the numerical results we obtain are identical to those obtained from Eqs. (8.9) and (8.13) when $k = 3$.

By applying Eq. (8.2) to the series, $\mathcal{T}_3^L(a)$ given by Eq. (7.7) with the first N_L terms have been removed, we obtain the exact terminant sum for the algebraic asymptotic expansion, which is

$$\Delta\mathcal{T}_3^L(a, N_L)_E = (-1)^{N_L+1} \frac{3^{6N_L+7/2} a^{2N_L+1}}{2^{6N_L+5} \pi^{6N_L+6}\, i} \int_{-1/4-i\infty}^{-1/4+i\infty} dt\; \eta^t\, \Gamma(1+t)$$
$$\times\; \Gamma(-t)\, \zeta(6N_L + 4 + 6t)\, \Gamma(2N_L + 4/3 + 2t)\, \Gamma(2N_L + 5/3 + 2t) \;. \qquad (8.14)$$

In the above equation $\eta = (3/2\pi)^6 a^2$ and we have carried out the sum over n in the original series by writing it in terms of the Riemann zeta function.

For $N_L = 0$, Eq. (8.14) simply yields the entire algebraic series $\mathcal{T}_3^L(a)$ and is identical to the general form presented in Eq. (8.9) for $k = 3$ if we recast the integrand using the reflection formulas for both the gamma and zeta functions, Nos. 8.334.2 and 9.535.2, respectively, in Ref. [16].

The complex integral in Eq. (8.14) can be readily evaluated to arbitrary precision by using a mathematical software package such as Mathematica [31] which incorporates fast algorithms for special functions of complex arguments. There are several reasons why it is more advantageous to use the terminant sum expression given by Eq. (8.14) over the terminant sum expression given by Eq. (7.18) for numerical computations. First, the computational effort required to evaluate a three-dimensional integral is significantly greater than that for a one-dimensional integral. Second, the integrand in Eq. (8.14) falls off very rapidly away from the real axis along the contour. E.g., for $a = 0.4$ and $N_L = 1$, a cut-off at $t = -1/2 \pm 3i$ already yields 8 digit precision and an accuracy of 10^{-12} for the integral. In this section, precision will refer to the number of significant figures while accuracy will refer to the size of the remainder. A cut-off at $t = -1/2 \pm 10i$ gives 35 digit precision and an accuracy of 10^{-39}. The integral can be simplified further since the real part of the integrand is symmetric along the contour moving away from the real axis, while the imaginary part is antisymmetric. Finally, Eq. (8.14) no longer contains the n-sum of Eq. (7.18) which may require the computation of many terms to achieve the desired level of accuracy. For example, to achieve an accuracy of 10^{-21} with $a = 0.4$ and $N_L = 1$ by means of Eq. (7.18), at least 78 terms were required to display the result given in Table 17.

The numerical expediency of the above approach is displayed in Table 19 where we give CPU times for the computation of the terminant sum $\Delta\mathcal{T}_3^L(a, N_L)_E$ via Eqs. (7.18) and (8.14) using machine-precision numbers. With arbitrary precision the computational effort increases greatly. For example, it took our computer system about half an hour of CPU time to evaluate $\Delta\mathcal{T}_3^L(0.4, N_L)_E$ to 17 significant figures, which appears in Table 15, whereas to evaluate it to 8 significant figures we can see from Table 19 that it only took 1.42 s. We see also from Table 19 that it takes 60712 s to evaluate $\Delta\mathcal{T}_3^L(a, N_L)_E$ to 8 significant figures by using Eq. (7.18) whereas by using Eq. (8.14) it only takes 1.42 s. The latter approach compares very favourably with the direct evaluation of $S_3(a)$ (0.73 s). Thus we see that the computational effort involved in evaluating the dominant asymptotic series by using Eq. (8.14) is now of the same order of magnitude as the direct evaluation of $S_3^L(a)$. Previously, Berry and Howls [6] found that standard numerical procedures were much superior in terms of speed using superasymptotics and hyperasymptotics, and consequently, stated that the principal reason for considering a hyperasymptotic solution was that it provides a testbed for studying in great detail the structure of a physical theory as an important parameter approaches a limiting value. We have shown, however, that Mellin-

Barnes regularisation provides us now with a technique that can make the superasymptotic solution compete with standard numerical techniques.

Application of Eq. (8.2) to the (subdominant) exponential series $\mathcal{T}_3^K(a)$, given by Eq. (7.20), after the first N_K terms have been removed, yields

$$\Delta\mathcal{T}_3^K(a,N_K)_E = \mathrm{Re}\left[\frac{e^{-i(5\pi/8+3\pi N_K/4)}}{2^{2N_K}(6\pi a)^{1/4}\sqrt{\pi}\,\Gamma(1/6)\Gamma(5/6)}\right.$$
$$\times \sum_{n=1}^{\infty}\frac{e^{\sqrt{2z}(i-1)}z^{-N_K/2}}{n^{1/4}}\int_{-1/10-i\infty}^{-1/10+i\infty}dt\left(\frac{e^{i\pi/4}}{4\sqrt{z}}\right)^t\Gamma(1+t)\Gamma(-t)$$
$$\left.\times\frac{\Gamma(N_K+1/6+t)}{\Gamma(N_K+1+t)}\Gamma(N_K+5/6+t)\right] , \tag{8.15}$$

where $z=(2n\pi/3)^3a^{-1}$ and we have chosen the contour of integration so that Eq. (8.15) holds for $N_K=0$ as well. In this special case, Eq. (8.15) represents the entire original series $\mathcal{T}_3^K(a)$. Even though we have not been able to show analytically that the expression for $\mathcal{T}_3^K(a)$ derived from Eq. (8.15) and our general expression given by Eq. (8.13) for $k=3$ are identical, we have tested them numerically to 50 figures precision without finding a discrepancy.

For the same reasons as mentioned above, the terminant sum for the exponential series Eq. (8.15) is also superior to its corresponding convergent integral form Eq. (7.26), but here the gain in numerical expediency is not as great as in the previous example because the latter only possesses a double integral as opposed to the former's triple integral. In addition, the integrand in Eq. (8.15) does not fall off as quickly away from the the real axis as the integrand for the algebraic terminant sum in Eq. (8.14) does and it does not have the special symmetry of the contour about the real axis. The sum over n in Eq. (8.15), however, does not present a problem since the exponential factors in the sum converge extremely quickly, in marked contrast to the n-sum for the algebraic series. Thus, for $a=0.4$, we find that only two terms are required from the n-sum to evaluate the terminant sum for the exponential series whereas 78 terms are required to evaluate the terminant sum for the algebraic series.

The numerical expediency of the above approach is again illustrated in Table 19 where we list CPU times for the computation of the terminant sum, $\Delta\mathcal{T}_3^K(a,N_K)_E$, via Eqs. (7.26) and (8.15) using machine-precision numbers. We see that there is still a clear advantage in using the Mellin-Barnes regularisation approach which increases further as higher precision is required. In order to obtain the 17 figure value for $\Delta\mathcal{T}_3^K(0.4,N_K)_E$ given in Table 17 , it took our computer system about 80 s of CPU time using Eq. (8.15).

C. Terminant sums for $S_{p/q}(a)$

At the beginning of this section, we derived exact analytical expressions for the algebraic series, $\mathcal{T}_k^L(a)$, and the exponential series, $\mathcal{T}_k^K(a)$, of the generalised Euler-Jacobi series in terms of Mellin-Barnes line integrals. The original asymptotic expansions which these integrals represent can be readily recovered simply by shifting the line contour to the right and evaluating the residues of the poles crossed in the process. The exact terminant sums $\Delta\mathcal{T}_k^L(a, N_K)_E$ and $\Delta\mathcal{T}_k^K(a, N_K)_E$ are then given by their corresponding integrals and evaluated by shifting their line contours to the right.

By shifting the line contour in Eq. (8.9) for the algebraic series $\mathcal{T}_k^L(a)$ to the right by $2N_L$, we cross N_L poles and thus obtain the first N_L terms of the algebraic expansion. The exact terminant sum which corresponds to breaking off the expansion after N_L terms therefore becomes

$$\Delta\mathcal{T}_k^L(a, N_L)_E = \frac{(-1)^{(k-1)/2}}{2\pi i} \int_{1/2+2N_L-i\infty}^{1/2+2N_L+i\infty} dt \, \frac{a^t\,\zeta(-kt)\Gamma(-t)\sin(\pi t/2)}{\sin(k\pi t/2)} . \tag{8.16}$$

For $N_L = 0$, we recover the entire algebraic series, $\mathcal{T}_k^L(a)$, while for $k = 3$, Eq. (8.16) is identical to Eq. (8.14) for the terminant sum, $\Delta\mathcal{T}_3^K(a, N_K)_E$. An algebraic series and corresponding terminant sum only exist when k is odd, since all integrals vanish for even k.

In order to obtain an asymptotic expansion for the exponential series we cannot use the integral representation Eq. (8.13) for $\mathcal{T}_k^K(a)$ since all the residues of its poles are zero. However, by performing the change of variable $s = t/2 + 1/4$ in the integral representation for the algebraic series Eq. (8.9) and then subtracting it from the tail of the generalised Euler-Jacobi series Eq. (8.5), we get the alternative expression for the exponential series of

$$\mathcal{T}_k^K(a) = \frac{1}{2\pi i} \int_{1/2-i\infty}^{1/2+i\infty} dt \left\{ a^t\zeta(-kt)\Gamma(-t) - (-1)^{(k-1)/2} a^{2t-1/2} \right.$$
$$\left. \times\ \zeta\left(-k(2t-1/2)\right)\Gamma\left(-(2t-1/2)\right) \frac{\sin\left(\pi(t-1/4)\right)}{\sin\left(k\pi(t-1/4)\right)} \right\} . \tag{8.17}$$

By using this result, we can now obtain an asymptotic expansion. By shifting the line contour in Eq. (8.17) to the right by N_K, we cross N_K poles and thus, obtain the first N_K terms of the subdominant expansion. The exact terminant sum which corresponds to breaking off the expansion after N_K terms becomes

$$\Delta\mathcal{T}_k^K(a, N_K)'_E = \frac{1}{2\pi i} \int_{1/2+N_K-i\infty}^{1/2+N_K+i\infty} dt \left\{ a^t\,\zeta(-kt)\Gamma(-t) - (-1)^{(k-1)/2} \right.$$
$$\left. \times\ \frac{\sin(\pi(t-1/4))}{\sin(k\pi(t-1/4))} \right\} a^{2t-1/2}\,\zeta\left(-k(2t-1/2)\right)\Gamma\left(-(2t-1/2)\right) . \tag{8.18}$$

Two remarks are necessary with regard to Eq. (8.18). First, the subdominant series $\mathcal{T}_k^K(a)$ is represented by $N_K = 0$ and this constitutes the entire subdominant contribution to the tail of the generalised Euler-Jacobi series. Although it is not a priori clear that this is identical to the exponential series $\mathcal{T}_k^K(a)$, obtained via Luke's asymptotics in the preceding sections, we have not been able to detect any numerical discrepancy in the cases considered. This justifies the use of the same symbol $\mathcal{T}_k^K(a)$ for both series. Second, term by term, the asymptotic expansion corresponding to $\Delta\mathcal{T}_k^K(a, N_K)'_E$ is different from the asymptotic expansion corresponding to $\Delta\mathcal{T}_k^K(a, N_K)_E$ in the previous section, even though both expansions recover the entire exponential series $\mathcal{T}_k^K(a)$. Hence, $\Delta\mathcal{T}_k^K(a, N_K)'_E$ is different from $\Delta\mathcal{T}_k^K(a, N_K)_E$.

Unfortunately, we do not have a general form for the asymptotic expansion for $\mathcal{T}_k^K(a)$ used in the previous section. However, it is still possible to study $\mathcal{T}_k^K(a)$ individually for each k as we have done in Sec. 6 and then to derive the corresponding exact terminant sums $\Delta\mathcal{T}_k^K(a, N_K)_E$. Thus, by direct application of the regularisation formula Eq. (8.2) to the subdominant series in Eqs. (7.20), (6.30), (6.105), (6.48), and (6.129), we can derive a general expression for the subdominant terminant sums $\Delta\mathcal{T}_k^K(a, N_K)_E$ for k =3, 4, 5, 6 and 7. Thus, we find

$$\Delta\mathcal{T}_k^K(a, N_K)_E = \mathcal{C}_1 \cdot \mathrm{Re}\left[\frac{\exp(-i\pi[1+\gamma_1(N_K-1/2)])}{\pi^{(2k-3)/(2k-2)}a^{1/(2k-2)}}\right.$$
$$\times \sum_{n=1}^{\infty} \frac{z^{-N_K/(k-1)}}{n^{(k-2)/(2k-2)}} \exp(-(k-1)z^{1/(k-1)}e^{i\gamma_1\pi}) \int_{c-i\infty}^{c+i\infty} dt\, N_{N_K+t}\, \Gamma(-t)$$
$$\left.\times\ \Gamma(1+t)\left(\frac{e^{i(1-\gamma_1)\pi}}{z^{1/(k-1)}}\right)^t\right] + \mathcal{C}_2 \cdot \mathrm{Re}\left[\frac{\exp(-i\pi[1+\gamma_2(N_K-1/2)])}{\pi^{(2k-3)/(2k-2)}a^{1/(2k-2)}}\right.$$
$$\times \sum_{n=1}^{\infty} \frac{z^{-N_K/(k-1)}}{n^{(k-2)/(2k-2)}} \exp(-(k-1)z^{1/(k-1)}e^{i\gamma_2\pi}) \int_{c-i\infty}^{c+i\infty} dt\, N_{N_K+t}$$
$$\left.\times\ \Gamma(1+t)\,\Gamma(-t)\left(\frac{e^{i(1-\gamma_2)\pi}}{z^{1/(k-1)}}\right)^t\right], \tag{8.19}$$

where $\gamma_1 = k/(2k-2)$, $z = (2n\pi/k)^k a^{-1}$ and the values for the parameters $\mathcal{C}_1$, $\mathcal{C}_2$, and γ_2 are listed in Table 20.

Eq. (8.19) becomes useful only once an analytic expression for the N_r has been obtained as a function of r. As indicated in Sec. 6, the N_r are the solutions of the various recursion relations that arise when studying different p/q values for the generalised Euler-Jacobi series. For $k = 3$, viz. $p/q = 3$ in the generalised Euler-Jacobi series, the N_r can be solved exactly as given by Eq. (6.89). Introducing this result into Eq. (8.19) then gives Eq. (8.15) for the terminant sum $\Delta\mathcal{T}_3^K(a, N_K)_E$. For $k > 3$, we have described in Sec. 7 how to evaluate accurate asymptotic forms for N_r.

Whilst we have not investigated $\mathcal{T}_k^K(a)$, and hence $\Delta\mathcal{T}_k^K(a, N_K)_E$, for $k > 7$, we expect that much of the fundamental structure presented in Eq. (8.19) will be retained. That is, it would be particularly interesting to examine whether the extra subdominant series that emerge as k increases, simply yield similar terms to those appearing in Eq. (8.19). We could represent the extra terms by the parameters $\{\mathcal{C}_3, \gamma_3\}$, $\{\mathcal{C}_4, \gamma_4\}$, etc. Then a general pattern for the $\mathcal{C}_i$, γ_i may evolve in a similar manner to γ_1.

9. CONCLUSION

The principal aims in this work were to produce an exact expression for the series given by Eq. (1.3), which we have called the generalised Euler-Jacobi series, and then to use our exact expression given by Eq. (2.14) or by Eq. (2.16) for $p \geq q+1$ to evaluate the small a-asymptotics of the series. We have referred to either result as the generalised Euler-Jacobi inversion formula for $S_{p/q}(a)$. During the course of this work, we found that the generalised Euler-Jacobi series was itself a special case of a more general series, which we have denoted by $S^{r,m}_{p/q}(a)$. As a consequence, we were able to develop an exact expression for the series $S^{r}_{p/q}(a)$ (see Eq. (1.4)), which has been studied by Berndt [1]. Each of the exact expressions for the series described here can be expressed either in hypergeometric functions or in Meijer G-functions. Thus, whilst we have been concerned mainly with evaluating the small a-asymptotics of the generalised Euler-Jacobi series, much of the asymptotic analysis presented in Secs. 3, 5 and 6 can be applied to the other series. Aside from presenting exact expressions for the generalised Euler-Jacobi series and the more general versions akin to it, we also showed in Sec. 2 how these exact expressions could be used to develop exact expressions for even more esoteric series such as that given in Eq. (2.28).

In Sec. 3 we examined the general asymptotic behaviour for small a of the generalised Euler-Jacobi series by first examining the large z-asymptotics of the hypergeometric functions ${}_{q+1}F_p$ appearing in our inversion formula. Using the results given in Luke [8], we indicated that two distinct contributions to the asymptotic forms of the hypergeometric functions exist, which had to be considered in evaluating the final asymptotic forms for the generalised Euler-Jacobi series. The first of these contributions consisted of oscillatory exponentials, which we referred to as the K-asymptotics, whilst the second contribution referred to as the L-asymptotic contribution consisted only of algebraic powers of a.

We mentioned that the dominant terms arising from the K-asymptotics of each hypergeometric function in our inversion formula consisted of a growing exponential multiplied by a series in powers of z. In fact, we showed for a hypergeometric function ${}_2F_p$ that the number of growing exponentials coming from the K-asymptotics increased depending on the number of positive integers k obeying the condition $(2k+1) < (p-1)/2$ with $p \geq 3$. Since the generalised Euler-Jacobi series is a convergent series, we pointed out that there must be a unique cancellation of these growing exponentials in our inversion formula. Thus for values of p/q other than even integers, the L-asymptotics became the dominant expressions of the asymptotic forms for the generalised Euler-Jacobi series, while the K-asymptotic contributions were relegated to the role of being the subdominant terms. However, for the cancellation to occur, we noted that the coefficients N_r and powers of the algebraic series multiplying the oscillatory exponential had to be identical. If this were the case, then cancellation could occur by combining all the oscillations for each growing exponential. This, in turn, meant that the recursion relations used to determine the N_r for each hypergeometric function in our inversion formula had to be identical even though the recursion relations are dependent on the indices or parameters of a hypergeometric function. Hypergeometric functions with the same recursion relation will have the same K-asymptotics except for an algebraic power of their variable and a slight difference in phase appearing in their oscillatory terms. Hence, even though the indices or parameters of each hypergeometric function in the inversion formula were different, the recursion relation generated by them had to be the same for each hypergeometric function. Thus we found in Sec. 6 that the hypergeometric functions for each value of p/q in our inversion formula belong to special families yielding the same coefficients N_r in their K-asymptotics. We noted also that a pattern exists to obtain the recursion relations of higher order hypergeometric functions belonging to a family. For example, we were able to predict the recursion relations for ${}_1F_{k-1}$ and ${}_2F_k$ hypergeometric functions from the ${}_0F_{k-2}$ hypergeometric functions appearing in $S_k(a)$. By permuting the ω_j values in the $T(t)$, we indicated in Sec. 3 that all the hypergeometric functions with the same N_r coefficients in their K-asymptotics could be determined. This determination of families of hypergeometric functions with the same N_r coefficients is probably a consequence of the fact that successive upper and lower indices or parameters appearing in the hypergeometric functions of our inversion formula differ by constant values; for the upper parameters it is q^{-1} whereas for the lower parameters it is p^{-1}. Since the parameters of the hypergeometric functions in the other series results given in Sec. 2 display the same behaviour, we expect that similar patterns regarding the recursion relations for the hypergeometric functions in these results. Therefore, given an arbitrary hypergeometric function ${}_pF_q((\alpha_p);(\beta_q);z)$, it should be possible for us to determine the other ${}_pF_q$ hypergeometric functions with the same N_r coefficients by permuting each of the ω_j appearing in the $T(t)$ and then evaluating the λ_j using the equation given in Sec. 2. Hence, the whole group of ${}_pF_q$ hypergeometric functions with the same N_r coefficients in their K-asymptotics can be determined. If the upper and lower parameters do not differ by constant values, then determination of higher order hypergeometric functions with the same N_r coefficients may be extremely difficult.

As a result of the cancellation of the growing exponential terms, we stated that only the decaying oscillatory exponentials from the K-asymptotics would remain. That is, only the subdominant K-asymptotic contributions in each of the hypergeometric functions appearing in the inversion formula would contribute to the final asymptotic form for the generalised Euler-Jacobi series. These were, therefore, the exponentially decaying terms missing in the Ramanujan-Berndt result, Eqs. (2.2) and (2.3), which gives the small a-asymptotics for the generalised Euler-Jacobi series only in terms of a zeta function series in algebraic powers of a. As we explained in the introduction, it had already been found that $S_2(a)$ could be expressed in terms of the exponentials given by Eq. (1.2), whereas the zeta function series in the Ramanujan-Berndt result yielded zero.

We also showed in Sec. 3 that there was no contribution from the L-asymptotics of the generalised Euler-Jacobi series for p/q equal to an even integer. This occurred because all the hypergeometric functions for $S_{2k}(a)$ could be reduced to ${}_0F_{2k-2}$ hypergeometric functions, which do not possess L-asymptotic contributions. Hence, only decaying oscillatory exponentials arising from the K-asymptotics for these hypergeometric functions contributed to the final asymptotic forms for $S_{2k}(a)$. For $p/q = 2k + 1$, however, we showed that there would be an L-asymptotic contribution because one of the hypergeometric functions in $S_{2k+1}(a)$ could only reduce to a ${}_1F_{2k}$ hypergeometric function. We then evaluated the L-asymptotics from this hypergeometric function, which resulted in a power series in a. By interchanging the summations, this power series could be expressed as the zeta series of the Ramanujan-Berndt result. Thus the asymptotic forms for $S_{2k+1}(a)$ were expected to consist of a zeta series in powers of a and decaying oscillatory exponential terms.

In Sec. 4 we applied the method of steepest descent to evaluate the asymptotics of $S_{p/q}(a)$ and its more general relative $S^r_{p/q}(a)$ for p/q equal to 3. For this case we found that the path of steepest descent intercepted the y-axis, so that both the contribution from the vicinity of the saddle point and the contribution from the imaginary axis to the intercept formed the total asymptotics for $S_3(a)$. Evaluation of the saddle point contribution to leading order yielded the decaying oscillatory exponential term whilst evaluation of the integral along the imaginary axis to leading order yielded the zeta series of the Ramanujan-Berndt result. Although the complete series arising from the saddle point can be obtained (usually after much tedious work), our main aim was to use the results in Sec. 4, namely Eqs. (4.12) and (4.16), as a check on those presented in Sec. 6, where the $S_3(a)$ series would be studied. Furthermore, we showed for higher integer values of p/q that selecting the appropriate saddle point and concomitant path of steepest descent became increasingly difficult.

In Sec. 5 we examined the small a-expansions for the generalised Euler-Jacobi series for values of p/q less than 2. Since the $p/q = 1$ case was well-known, we examined this case by first evaluating the asymptotics for $S^r_1(a)$

and then considering $r = 1$. For values of $p/q < 1$, only the representation of $S_{p/q}(a)$ in terms of hypergeometric functions, i.e. Eq. (2.14), could be used to determine the small a-series. In this section we were able to develop the a-series for p/q equal to 1/3, 1/2 and 2/3, which are given by Eqs. (5.8), (5.15) and (5.22), respectively. We found that for $p/q = 1/3$, the result could be expressed in terms of Lommel functions; for $p/q = 1/2$ the result could be expressed either in Fresnel integrals or in Bessel and Anger-Weber functions all of order 1/2, while the result for $p/q = 2/3$ could be expressed in terms of confluent hypergeometric functions. After examining the $p/q = 1$ case, we considered p/q equal to 4/3 and 3/2. For $p/q > 1$ the expansions for $S_{p/q}(a)$ are asymptotic. Although the $p/q = 3/2$ case could be partially expressed in special Bessel functions of order 2/3 (see Eq. (5.70)), we had to utilise Luke's theory [8] given in Sec. 3 to develop the complete asymptotics. Here we encountered for the first time the Stokes phenomenon which was responsible for yielding alternative asymptotic expansions for the hypergeometric functions in $S_{4/3}(a)$ and $S_{3/2}(a)$. However, we were able to develop the asymptotics since only one representation for each case yielded a real result. From our analysis in Sec. 5, we found that the small a-asymptotics for the generalised Euler-Jacobi series for $p/q < 2$ could be eventually written in the form given by the Ramanujan-Berndt result. Thus for $p/q < 2$, there were no exponential corrections missing in Eq. (2.2).

We began our study of integer cases ($q = 1$) greater than or equal to 2 in Sec. 6 by considering even integers up to $p = 6$. Since the $p = 2$ case has been well-known for over 150 years, we examined this case by evaluating $S_2^r(a)$ for some values of r given by Eqs. (6.8) and (6.9). We were able to express our result for $S_2^r(a)$ in terms of ${}_1F_1$ hypergeometric functions which for $r=1$, i.e. the generalised Euler-Jacobi series, reduced easily to the exponentials appearing in Eq. (1.2). For $r \neq 1$ we were able to express $S_2^r(a)$ in terms of parabolic cylinder functions given by Eq. (6.7), whose asymptotics are also affected by the Stokes phenomenon. For the cases of p/q equal to 4 and 6, we found that $S_4(a)$ and $S_6(a)$ could be expressed in terms of ${}_0F_2$ and ${}_2F_4$ hypergeometric functions respectively, thereby confirming that no L-asymptotic contribution would exist for these cases. Furthermore, the dominant K-asymptotic contributions for both cases cancelled as indicated in Sec. 3. Hence, we were left with series consisting of decaying oscillatory exponentials given by Eqs. (6.30) and (6.48). As expected, the leading order term of Eq. (6.30) agreed with the leading order term from the vicinity of the saddle point for $S_4(a)$, i.e. Eq. (4.16).

To complete our study of integer cases for the generalised Euler-Jacobi series, we then looked at the odd integers 3, 5, and 7. We were able to express $S_3(a)$ in special Bessel functions of order 1/3 given by Eq. (6.62). This was useful since we could use existing asymptotic forms for Bessel functions to make a comparison with the asymptotics obtained by using Luke's theory. We found that the asymptotic form for $S_3(a)$ using our generalised Euler-Jacobi inversion formula could be expressed in terms of two ${}_0F_1$ and one ${}_1F_2$ hypergeometric functions. As indicated in Sec. 3, the L-asymptotics for the ${}_1F_2$ produced the Ramanujan-Berndt result, which could be traced to part of the asymptotics appearing in the Anger functions in Eq. (6.62). We also found that the dominant terms in the K-asymptotic contribution cancelled because the coefficients N_r for the K-asymptotics of each hypergeometric function obeyed the same recursion relation even though the parameters or indices of the hypergeometric functions were different. We were then able to show that the remaining subdominant terms could be combined to produce the same asymptotic form obtained by using the known asymptotics for Bessel functions. Furthermore, the resultant leading order terms of the K-asymptotic contribution and the L-asymptotic contribution combined to yield the leading terms of the asymptotic form we obtained for $S_3(a)$ by using the method of steepest descent (compare Eq. (6.82) to Eq. (4.12)).

The asymptotic forms of the generalised Euler-Jacobi series for $p = 5$ and $p = 7$ are given by Eqs. (6.105) and (6.130), respectively. In general, evaluation of the generalised Euler-Jacobi series is more difficult for the odd integer cases than for the even integer cases because the sum over hypergeometric functions or Meijer G-functions reduces to simpler forms for even integers than for adjacent odd integers. For example, $S_6(a)$ was eventually expressed in terms of three ${}_0F_4$ hypergeometric functions given in Eq. (6.39) whereas for $p = 5$, the expression for $S_5(a)$ was eventually expressed in terms of four ${}_0F_3$ hypergeometric functions and one ${}_1F_4$ hypergeometric function.

It is particularly interesting to observe for $p = 6$ and $p = 7$ that additional subdominant terms emerge from the K-asymptotic contributions for the first time. As p or β_0 increases, we see from Eq. (3.1) that the number of terms in both sums increases although the complexity in evaluating higher integers of the gamma functions can be reduced by selecting a value for r such that $r - 1$ is close to $\beta_0 - r - 1$. For example, in the $p = 7$ case where $\beta_0 = 6$, we chose r equal to 3, which meant that only $\Gamma_p^{1,q}(2)$ were required. In general, by combining the two sums in Eq. (3.1), we found that the exponentials in the K-asymptotics all contained the factor $\exp(\beta_0 z^{1/\beta_0} \cos((2m+1)\pi/2\beta_0))$. Thus, because more integers m satisfied $-1 \leq \cos((2m+1)\pi/2\beta_0) < 0$, we obtained for the first time additional subdominant terms in the final asymptotic forms of the generalised Euler-Jacobi series for $p = 6$ and $p = 7$, not obtained for lower p values.

It can now be seen that as p increases further, the number of subdominant terms appearing in the asymptotic forms for the generalised Euler-Jacobi series will also increase. This means that more and more $\Gamma_p^{1,q}(k)$ and their complex conjugates will have to be evaluated. Determining these quantities for the various hypergeometric functions appearing in $S_p(a)$ becomes increasingly onerous as p increases but can be made less tedious by using an appropriate trigonometric identity for the case in hand. For example, we utilised Eq. (6.121) for $p = 7$. For more details regarding the evaluation of the $\Gamma_p^{1,q}(k)$, we refer the reader to p. 196 of Luke [8], who shows how to evaluate them for higher values of k beyond $k = 2$.

To determine the K-asymptotics of higher values at p/q for the generalised Euler-Jacobi series or for that matter any of the other series presented in this work, the parameter β_0 for the hypergeometric functions or Meijer G-functions appearing in the inversion formulae of Eqs. (2.15) and (2.16) must be found and the value of r for the summation in Eq. (3.2) should be chosen so that $r - 1$ is as close as possible to $\beta_0 - r - 1$. Then all values of k in the sums yielding growing exponential terms can be excluded. The recursion relation for only one of the hypergeometric functions need be evaluated to obtain the N_r coefficients of all hypergeometric functions. As mentioned in the previous paragraph, the $\Gamma_p^{1,q}(k)$ in Eq. (3.1) should be determined according to the prescription given in Luke [8]. These may be simplified by using a trigonometric identity obtained by summing a finite geometric series such as Eq. (6.104). The resulting K-asymptotic expression may be simplified by combining all exponentials.

The L-asymptotics do not contribute if the asymptotics yield only hypergeometric functions of the form ${}_0F_k$, which we have seen occurs when $p = 2k$ for the generalised Euler-Jacobi series. As can be seen from Eq. (3.3) the L-asymptotics produce a series in inverse powers of z, which means powers of a. We found that the L-asymptotics can be simplified by applying Gauss' multiplication formula and that interchanging the summations produces the zeta series as given by Eq. (3.27).

In Sec. 7 we applied the newly emerging subject of asymptotics beyond all orders to the asymptotic expansion for $S_3(a)$ obtained from our inversion formula. Here, we truncated both asymptotic series in this expansion given by Eq. (6.82) and then applied Dingle's theory of terminants to evaluate the remainders of both series, which were given ultimately in convergent integral representations. We used two methods to evaluate the expressions of the remainders in terms of sums of terminants, which we referred to as terminant sums. The first method was that used by Berry [5] and was referred to as the 'asymptotics of asymptotics' approach. Basically, in this method one aims at showing that the late terms of an asymptotic series have the universal behaviour of $\Gamma(k + \text{const.})/(\text{variable})^k$ for large k, so that the series has the general form of a terminant given on pp. 406-407 of Ref. [3]. We showed that for small values of a this was quite accurate when compared with the exact approach but deteriorated when a became greater than 0.01. In the exact approach we utilised the integral representation for the beta function to express the remaining terms of each truncated asymptotic series in Eq. (6.82) in terms of terminant sums. A particularly nice feature of this method is that one does not need to truncate asymptotic series near their optimal number of terms or for very large values of k to evaluate their remainders. The terminant sums for the algebraic and exponential series appearing in the asymptotic expansion of $S_3(a)$ using the first approach appear in convergent integral representations respectively as Eqs. (7.16) and (7.25) while those evaluated by the exact approach appear as Eqs. (7.18) and (7.26).

In discussing the asymptotics of $S_3(a)$, we introduced a quantity called the tail, which we denoted by $\mathcal{T}_3(a)$ and which we set equal to $S_3(a) - \Gamma(1/3)/3a^{1/3} - 1/2$. We found for a equal to 0.01, 0.2, 0.4, 1 and 10 that when we combined both terminant sums using the exact approach with the truncated asymptotic series, we obtained the exact value of $\mathcal{T}_3(a)$ to an extremely high level of accuracy ($\sim 10^{-57}$ for a equal to 0.01 and $\sim 10^{-15}$ for a equal to 10). Our level of accuracy went well beyond the hyperasymptotic accuracy being put forward by Berry [5]. Although we believe that the asymptotic expansion for $S_3(a)$ is exact, we have not conclusively proven that there are no more highly transcendental exponentially decaying series missing in Eq. (6.82). However, we have shown that the study of asymptotics beyond all orders opens up a new frontier in both applied mathematics and theoretical physics, that is for the first time the heretofore unsolved intermediate region can now be explored.

We also showed in Sec. 7 how to obtain extremely accurate asymptotic values of the subdominant exponential series when one is unable to solve the recursion relation exactly. Again we relied on the universal behaviour of the late terms of these series to develop an elegant iterative technique, which we applied to $S_4(a)$. We concluded the section by showing how our exact approach for terminating the algebraic series in $S_3(a)$ could be applied to $S_5(a)$ and $S_7(a)$ whose terminant sums appear as Eqs. (7.45) and (7.46) respectively. As can be seen from these results, the dimensionality of the resulting integrals increases as p/q increases, thereby making them less attractive for numerical computation.

Even though the forms we gave for the terminant sums of $S_3(a)$ in Sec. 7 were amenable to numerical computation, we found that when using Mathematica [31] to evaluate the various multi-dimensional integrals we could not achieve the accuracy needed for our study in a time-expedient manner, as described in Sec. 8. Thus, we devised the technique of Mellin-Barnes regularisation to represent the various terminant sums as Mellin-Barnes integrals, which enabled us to evaluate them to any level of accuracy we required. We were able to generalise our procedure to all odd integers and hence, can now evaluate the terminant sums for $S_5(a)$ and $S_7(a)$ instead of using their complicated convergent integral representations given in Sec. 7. Consequently, we were able to provide an integral representation for the exponentially decaying terms missing in the Ramanujan-Berndt result for the generalised Euler-Jacobi series.

We should mention that the inversion formulae given by Eqs. (2.13) and

(2.14) can be easily extended to the complex a-plane by noting that Eq. (2.11) is valid for complex a as long as Re $a >$ Im $\beta = 0$. If $a = a_1 + ia_2$, then we can replace a^q by $(a_1^2 + a_2^2)^{q/2} \exp(iq \arg a)$ in Eqs. (2.13) and (2.14). Hence, the inversion formula given by Eq. (2.14) can be modified to

$$S_{p/q}(a) = \frac{q\Gamma(q/p)e^{-iq\theta/p}}{p\,(a_1^2 + a_2^2)^{q/2p}} + \frac{1}{2} + \frac{(2\pi)^{(p-q)/2}}{(a_1^2 + a_2^2)^{q/2p}} \frac{q^{q/p+1/2}}{p^{3/2}} e^{-iq\theta/p}$$
$$\times \sum_{n=1}^{\infty} \sum_{l=0}^{p-1} (-1)^l \, (z')^{2l/p} \; e^{-2li\theta/p} \prod_{\Gamma}^{q,p} \Big((2l+1)/p, 0\Big)$$
$$\times \; {}_qF_{p-1}\Big(1, \Delta_{q-2}\big(q, \tau(1,0)\big); \Delta_{p-2}\big(p, 2l+2\big); \pm z' e^{-iq\theta + i\pi p/2 + i2s\pi}\Big)^{+} \quad , \qquad (9.1)$$

where $z' = (q/(a_1^2+a_2^2)^{1/2})^q (2n\pi/p)^p$ and $\theta = \arg a$. Although the asymptotics will now depend on the values of a_1 and a_2, we note that the exact values for the generalised Euler-Jacobi series will be altered by the phase of the variable in the ${}_qF_{p-1}$ hypergeometric function in Eq. (9.1).

It is most fitting as we begin to close this discourse that we return to Euler and Jacobi's theta function, $\theta_3(z)$, by studying Eq. (9.1) for $p/q = 2$. We now find

$$S_2(a_1 + ia_2) = \frac{\Gamma(1/2)\, e^{-i\theta/2}}{2(a_1^2 + a_2^2)^{1/4}} + \frac{1}{2} + \frac{\sqrt{\pi}\; e^{-i\theta/2}}{(a_1^2 + a_2^2)^{1/4}}$$
$$\times \sum_{n=1}^{\infty} \Big(\cosh\big(ze^{-i\theta}\big) - \sinh\big(ze^{-i\theta}\big)\Big) \quad , \qquad (9.2)$$

where $z = n^2\pi^2/(a_1^2 + a_2^2)^{1/2}$, which is, of course, the classic Jacobi theta function transformation [10,11].

If we put $p/q = 3$ into Eq. (9.1), then

$$S_3(a_1 + ia_2) = \frac{\Gamma(1/3)\, e^{-i\theta/3}}{3\,(a_1^2 + a_2^2)^{1/6}} + \frac{1}{2} + \frac{(2\pi)^{3/2} e^{-i\theta/2}}{9\,(a_1^2 + a_2^2)^{1/4}} \sum_{n=1}^{\infty} n^{1/2}$$
$$\times \Bigg[\sqrt{2}\Big\{\mathrm{ber}_{-1/3}\big(2\sqrt{z'}e^{-i\theta/2}\big) - \mathrm{bei}_{-1/3}\big(2\sqrt{z'}e^{-i\theta/2}\big)$$
$$+ \mathrm{bei}_{1/3}\big(2\sqrt{z'}e^{-i\theta/2}\big) - \mathrm{ber}_{1/3}\big(2\sqrt{z'}e^{-i\theta/2}\big)\Big\}$$
$$- \; e^{i\pi/4}\Big\{\mathbf{J}_{1/3}\big(2\sqrt{z'}e^{-i(\theta+\pi/2)/2}\big) - \mathbf{J}_{-1/3}\big(2\sqrt{z'}e^{-i(\theta+\pi/2)/2}\big)\Big\}$$
$$- \; e^{-i\pi/4}\Big\{\mathbf{J}_{1/3}\big(2\sqrt{z'}e^{-i(\theta-\pi/2)/2}\big) - \mathbf{J}_{-1/3}\big(2\sqrt{z'}e^{-i(\theta-\pi/2)/2}\big)\Big\}\Bigg] \quad , \qquad (9.3)$$

where $z' = (a_1^2 + a_2^2)^{-1/2}(2n\pi/3)^3$. The general result given by Eq. (9.1) and the specific values for p/q equal to 2 and 3 are presented above to show that all the results in this work have natural extensions to the complex a-plane.

We wish to point out that our analysis can be readily applied to alternating versions of the series considered here. The asymptotics of such series have been studied by Estrada and Kanwal [33], who, again, have obtained only the equivalent of the Ramanujan-Berndt result. Consequently, these authors have missed obtaining the corresponding exponential terms. To apply the analysis in Sec. 2 to their series

$$Q^r_{p/q}(a) = \sum_{n=0}^{\infty} (-1)^n n^{r-1} e^{-an^{p/q}} \ , \tag{9.4}$$

one merely splits it into two independent terms given by

$$Q^r_{p/q}(a) = 2^r S^r_{p/q}(2^{p/q}a) - S^r_{p/q}(a) \ , \tag{9.5}$$

whose exponential corrections follow from our detailed analysis.

In this work we have derived an exact inversion formula (Eq. (2.14) or Eq. (2.16)) for the generalised Euler-Jacobi series. From this result, we have been able to develop asymptotic expansions by applying the asymptotic theory for hypergeometric functions presented in Ch. 5 of Luke [8]. As outlined in Sec. 3, this theory includes subdominant terms not contained in the corresponding result obtained by Ramanujan and Berndt. We have shown that for even integers these subdominant terms become the major contribution to the asymptotic expansions whereas the Ramanujan-Berndt result is vacuous. In addition, we have been able to derive accurate expansions for the generalised Euler-Jacobi series, which have heretofore never been evaluated. Although the numerical analysis in Sec. 7 is consistent with Luke's statement that the asymptotic theory in Sec. 3 is complete, it would nevertheless be of interest to investigate analytically whether there are sub-subdominant terms in the K-asymptotics for the hypergeometric functions given by Eq. (2.14) or to understand why the asymptotic results given by Luke, who derived them from the seminal work of Braaksma [9], are complete. The evaluation of these hypothetical terms or the discovery of why Luke's results are complete is a frontier in the theory of asymptotics for hypergeometric functions and Meijer G-functions and will be addressed in the future. Only then will it be possible to give a fully rigorous presentation of the complete asymptotic structure for hypergeometric functions.

Finally, the results in Secs. 5, 7 and 8 have raised the possibility that a new theory of divergent series can be created that will enable asymptotics to break free from its shackles of inaccuracy and limitation in range of application to a discipline yielding precise results over previously unexplored regions. Of crucial importance to the creation of this theory will be the techniques of Borel summation and Mellin-Barnes regularisation, the latter technique enabling a much faster evaluation of the limits to divergent series than the application of standard numerical integration techniques to their convergent integral representations obtained via the former technique. However in Secs. 5, 7 and 8, we have been concerned only with series that alternate in phase, but to develop a theory of divergent series, we shall be required to consider series that do not alternate in phase. Whenever an asymptotic expansion contains a divergent series in which all its terms are homogeneous in phase and are of the same sign, it is an indication that a Stokes line/discontinuity has been encountered [3]. Therefore, any new theory of divergent series will need to take into account the Stokes phenomenon. This study is currently underway.

REFERENCES

[1] B.C. Berndt, *Ramanujan's Notebooks*, Part II (Springer-Verlag, New York, 1985), Ch. 15.

[2] H. Segur, S. Tanveer and H. Levine (eds.), *Asymptotics beyond All Orders* (Plenum Press, New York, 1991).

[3] R.B. Dingle, *Asymptotic Expansions; Their Derivation and Interpretation* (Academic Press, New York, 1973).

[4] F.W.J. Olver in *Asymptotic Solutions of Differential Equations and their Applications*, C.H. Wilcox (ed.) (Wiley, New York, 1964).

[5] M.V. Berry, *Asymptotics, superasymptotics and hyperasymptotics...*, in *Asymptotics beyond All Orders*, H. Segur, S. Tanveer and H. Levine (eds.) (Plenum Press, New York, 1991).

[6] M.V. Berry and C.J. Howls, *Hyperasymptotics*, Proc. R. Soc. Lond. **A 430**, 653-667 (1990).

[7] M.V. Berry and C.J. Howls, *Hyperasymptotics for integrals with saddles*, Proc. R. Soc. Lond. **A 434**, 657-675 (1991).

[8] Y.L. Luke, *The Special Functions and Their Approximations*, Vol. I (Academic Press, New York, 1969).

[9] B.L.J. Braaksma, *Asymptotic expansions and analytic continuations for a class of Barnes-integrals*, Compos. Math. **15**, pp. 239-341 (1964).

[10] R. Bellman, *A Brief Introduction to Theta Functions* (Holt, Rinehart & Winston, New York, 1961), p. 1.

[11] E.T. Whittaker and G.N. Watson, *A Course of Modern Analysis* (Cambridge University Press, Cambridge, 1988).

[12] K. Knopp, *Theory and Application of Infinite Series* (Blackie & Son Ltd., London, 1944), Ch. 14.

[13] G.H. Hardy, *Divergent Series* (Clarendon Press, Oxford, 1963).

[14] J.P. Ramis, *Théorèmes d'indices Gevrey pour les équations différentielles ordinaires*, Mem. Am. Math. Soc. **48**, 296 (1984).

[15] J. Thomann, *Resommation des séries formelles*, Numer. Math. **58**, 503-535 (1990).

[16] I.S. Gradshteyn and I.M. Ryzhik, *Tables of Integrals, Series and Products* (Academic Press, New York, 1982).

[17] M. Abramowitz and I.A. Stegun, *Handbook of Mathematical Functions* (Dover, New York, 1964).

[18] E.M. Wright, *A recursion formula for the coefficients in an asymptotic expansion*, Proc. Glasgow Math. Assoc. **4**, 38-41 (1958).

[19] P.M. Morse and H. Feshbach, *Methods of Theoretical Physics*, Vol. I (McGraw Hill, New York, 1982), p. 609.

[20] M.V. Berry, *Uniform asymptotic smoothing of Stokes' discontinuities*, Proc. R. Soc. Lond. **A 422**, 7-21 (1989).

[21] J.D. Murray, *Asymptotic Analysis* (Clarendon Press, Oxford, 1974).

[22] A.P. Prudnikov, Yu.A. Brychkov and O.I. Marichev, *Integrals and Series*, Vol. 3 (Gordon & Breach, New York, 1990).

[23] A.P. Prudnikov, Yu.A. Brychkov and O.I. Marichev, *Integrals and Series*, Vol. 1 (Gordon & Breach, New York, 1988).

[24] A. Apelblat, *Table of Definite and Indefinite Integrals* (Elsevier, Amsterdam, 1983).

[25] A.P. Prudnikov, Yu.A. Brychkov and O.I. Marichev, *Integrals and Series*, Vol. 2 (Gordon & Breach, New York, 1988).

[26] A. Erdelyi, W. Magnus, F. Oberhettinger & F.G. Tricomi, *Higher Transcendental Functions*, Vol. I (McGraw Hill, New York, 1953).

[27] G.N. Watson, *A Treatise on the Theory of Bessel Functions* (Cambridge University Press, Cambridge, 1944).

[28] F.W.J. Olver, *On Stokes' phenomenon and converging factors*, in *Proceedings of the International Symposium on Asymptotic and Computational Analysis* (Manitoba, Winnipeg 1989), R. Wong (ed.) (Marcel Dekker, New York, 1990), 329-355.

[29] W.G.C. Boyd, *Stieltjes transforms and the Stokes phenomenon*, Proc. Roy. Soc. Lond., **A 429**, 227-246 (1990).

[30] D.S. Jones, *Uniform asymptotic remainders*, in *Proceedings of the International Symposium on Asymptotic and Computational Analysis* (Manitoba, Winnipeg, 1989), R. Wong (ed.) (Marcel Dekker, New York, 1990) 241-264.

[31] S. Wolfram, *Mathematica*, 2nd Ed. (Addison-Wesley, Reading, Massachusetts, 1991).

[32] M.J. Lighthill, *Introduction to Fourier Analysis and Generalised Functions* (Cambridge University Press, Cambridge, 1975), Ch. 3.

[33] R. Estrada and R.P. Kanwal, *A distributional theory for asymptotic expansions*, Proc. R. Soc. Lond. **A 428**, 399-430 (1990).

TABLES

k	N_k (exact)	N_k (decimal)
0	1	$1.000\,000\,000\,0 \times 10^{0}$
1	$\frac{7}{144}$	$4.861\,111\,111 \times 10^{-2}$
2	$\frac{385}{41\,472}$	$9.283\,371\,913 \times 10^{-3}$
3	$\frac{39\,655}{17\,915\,904}$	$2.213\,396\,544 \times 10^{-3}$
4	$\frac{665\,665}{10\,319\,560\,704}$	$6.450\,516\,830 \times 10^{-5}$
5	$-\frac{1\,375\,739\,365}{1\,486\,016\,741\,376}$	$-9.257\,899\,50 \times 10^{-4}$
6	$-\frac{2\,053\,160\,864\,755}{1\,283\,918\,464\,548\,864}$	$-1.599\,136\,48 \times 10^{-3}$
7	$-\frac{400\,804\,002\,473\,875}{184\,884\,258\,895\,036\,416}$	$-2.167\,864\,39 \times 10^{-3}$
8	$-\frac{545\,523\,697\,484\,891\,125}{212\,986\,666\,247\,081\,951\,232}$	$-2.561\,304\,45 \times 10^{-3}$
9	$-\frac{639\,409\,620\,356\,437\,805\,875}{276\,030\,719\,456\,218\,208\,796\,672}$	$-2.316\,443\,69 \times 10^{-3}$
10	$-\frac{7\,400\,680\,096\,069\,804\,168\,625}{79\,496\,847\,203\,390\,844\,133\,441\,536}$	$-9.309\,400\,75 \times 10^{-5}$
11	$\frac{85\,225\,571\,098\,153\,435\,685\,610\,875}{11\,447\,545\,997\,288\,281\,555\,215\,581\,184}$	$7.444\,876\,929 \times 10^{-3}$
12	$\frac{548\,115\,663\,843\,414\,041\,224\,022\,298\,125}{19\,781\,359\,483\,314\,150\,527\,412\,524\,285\,952}$	$2.770\,869\,536 \times 10^{-2}$
13	$\frac{212\,014\,635\,165\,656\,643\,273\,521\,106\,914\,375}{2\,848\,515\,765\,597\,237\,675\,947\,403\,497\,177\,088}$	$7.442\,986\,193 \times 10^{-2}$
14	$\frac{133\,126\,972\,240\,163\,358\,184\,968\,745\,634\,504\,375}{820\,372\,540\,492\,004\,450\,672\,852\,207\,187\,001\,344}$	$1.622\,762\,411 \times 10^{-1}$
15	$\frac{90\,364\,081\,190\,288\,921\,441\,174\,372\,687\,896\,958\,125}{354\,400\,937\,492\,545\,922\,690\,672\,153\,504\,784\,580\,608}$	$2.549\,769\,812 \times 10^{-1}$
16	$\frac{9\,019\,416\,081\,298\,899\,889\,215\,820\,334\,616\,981\,356\,875}{816\,539\,759\,982\,825\,805\,879\,308\,641\,675\,023\,673\,720\,832}$	$1.104\,589\,944 \times 10^{-2}$
17	$-\frac{260\,021\,114\,680\,943\,473\,566\,117\,212\,676\,037\,766\,397\,671\,875}{117\,581\,725\,437\,526\,916\,046\,620\,444\,401\,203\,409\,015\,799\,808}$	$-2.211\,407\,544 \times 10^{0}$
18	$-\frac{3\,854\,975\,958\,092\,062\,993\,212\,896\,945\,123\,288\,990\,286\,762\,353\,125}{304\,771\,832\,334\,069\,766\,392\,840\,191\,887\,919\,236\,168\,953\,102\,336}$	$-1.264\,872\,783 \times 10^{1}$
19	$-\frac{2\,230\,702\,885\,075\,796\,330\,282\,978\,287\,686\,945\,999\,335\,191\,018\,440\,625}{43\,887\,143\,856\,106\,046\,360\,568\,987\,631\,860\,370\,008\,329\,246\,736\,384}$	$-5.082\,816\,262 \times 10^{1}$
20	$-\frac{4\,085\,674\,814\,497\,803\,591\,523\,660\,096\,191\,996\,942\,460\,008\,554\,035\,146\,875}{25\,278\,994\,861\,117\,082\,703\,687\,736\,875\,951\,573\,124\,797\,646\,120\,157\,184}$	$-1.616\,233\,096 \times 10^{2}$

TABLE 1. Values of N_k for $S_4(a)$

a	$S_3(a)$	$\mathcal{T}_3(a)$
10^{-12}	8 930 .295 115 692 492 103 852 309 803	$-8.333\,333\,333\,333\,333\,333\,33 \times 10^{-15}$
10^{-10}	1 924 .366 037 212 517 514 857 026 082	$-8.333\,333\,333\,333\,333\,333\,33 \times 10^{-13}$
10^{-8}	414 .984 372 954 398 710 769 710 859 9	$-8.333\,333\,333\,333\,333\,207\,07 \times 10^{-11}$
10^{-6}	89 .797 951 148 591 587 788 524 360 65	$-8.333\,333\,333\,332\,070\,707\,07 \times 10^{-9}$
10^{-4}	19 .738 659 538 791 851 411 173 251 77	$-8.333\,333\,320\,707\,071\,976\,45 \times 10^{-7}$
0.01	4.644 760 397 473 744 543 028 398 126	−0.000 083 332 071 075 898 002 043 80
0.05	2.923 502 824 440 646 091 459 826 882	−0.000 416 493 021 239 151 029 509 07
0.10	2.423 037 179 235 417 952 008 047 553	−0.000 828 857 977 100 396 179 988 38
0.20	2.025 146 612 801 709 916 375 693 083	−0.001 826 872 856 811 495 120 445 83
0.30	1.831 839 717 696 391 034 505 982 565	−0.002 094 489 545 186 296 258 210 78
0.40	1.711 102 649 525 038 553 041 641 717	−0.000 857 009 205 904 068 203 026 26
0.50	1.624 847 669 560 466 652 147 431 153	−0.000 236 014 190 472 851 783 736 96
0.60	1.557 041 475 279 054 828 173 337 839	−0.001 702 806 636 768 204 105 403 75
1.00	1.368 214 903 801 224 362 250 084 242	−0.024 764 607 768 024 858 967 680 07
2.00	1.135 335 395 771 787 411 153 117 541	−0.073 422 912 253 750 164 464 208 31
5.00	1.006 737 946 999 085 471 344 990 303	−0.015 479 639 338 743 717 485 311 58
10.0	1.000 045 399 929 762 484 851 535 591	0.085 561 026 975 280 440 748 491 398
100	1.000 000 000 000 000 000 000 000 000	0.307 613 396 278 748 165 181 196 405
1000	1.000 000 000 000 000 000 000 000 000	0.410 702 048 843 075 078 878 143 568
10^4	1.000 000 000 000 000 000 000 000 000	0.458 551 562 704 551 795 589 695 580
10^5	1.000 000 000 000 000 000 000 000 000	0.480 761 339 627 874 816 518 119 640
10^6	1.000 000 000 000 000 000 000 000 000	0.491 070 204 884 307 507 887 814 356
10^7	1.000 000 000 000 000 000 000 000 000	0.495 855 156 270 455 179 558 969 558

TABLE 2. Numerical values for a, $S_3(a)$ and $\mathcal{T}_3(a)$

N	$\mathcal{T}_3^L(0.01, N)$	R_N
1	−0.000 083 333 333 333 333 400 000 000	0.000 000 001 262 626 262 626 263 000 0
2	−0.000 083 332 070 707 070 707 070 707	0.000 000 000 000 369 383 169 934 640 5
3	−0.000 083 332 071 076 453 877 005 347	0.000 000 000 000 000 558 452 668 507 0
4	−0.000 083 332 071 075 895 424 336 840	0.000 000 000 000 000 002 686 664 527 8
5	−0.000 083 332 071 075 898 111 001 368	0.000 000 000 000 000 000 031 655 154 7
6	−0.000 083 332 071 075 898 079 346 213	0.000 000 000 000 000 000 000 774 711 5
7	−0.000 083 332 071 075 898 080 120 925	0.000 000 000 000 000 000 000 035 161 5
8	−0.000 083 332 071 075 898 080 120 925	0.000 000 000 000 000 000 000 002 724 2
9	−0.000 083 332 071 075 898 080 088 487	0.000 000 000 000 000 000 000 000 338 2
10	−0.000 083 332 071 075 898 080 088 149	0.000 000 000 000 000 000 000 000 064 0
11	−0.000 083 332 071 075 898 080 088 213	0.000 000 000 000 000 000 000 000 017 7
12	−0.000 083 332 071 075 898 080 088 195	0.000 000 000 000 000 000 000 000 006 9
13	−0.000 083 332 071 075 898 080 088 202	0.000 000 000 000 000 000 000 000 003 7
14	−0.000 083 332 071 075 898 080 088 199	0.000 000 000 000 000 000 000 000 002 7
15	−0.000 083 332 071 075 898 080 088 201	0.000 000 000 000 000 000 000 000 002 6
16	−0.000 083 332 071 075 898 080 088 197	0.000 000 000 000 000 000 000 000 003 2
17	−0.000 083 332 071 075 898 080 088 202	0.000 000 000 000 000 000 000 000 005 1
18	−0.000 083 332 071 075 898 080 088 197	0.000 000 000 000 000 000 000 000 010 2
19	−0.000 083 332 071 075 898 080 088 259	0.000 000 000 000 000 000 000 000 025 4
20	−0.000 083 332 071 075 898 080 088 182	0.000 000 000 000 000 000 000 000 077 0
21	−0.000 083 332 071 075 898 080 088 259	0.000 000 000 000 000 000 000 000 283 9
22	−0.000 083 332 071 075 898 080 082 550	0.000 000 000 000 000 000 000 001 260 6

TABLE 3. Numerical values for the algebraic series and its remainder with $a = 0.01$

N	$\mathcal{T}_3^L(0.2, N)$	R_N
1	−0.001 666 666 666 666 666 666 666 666	0.000 010 101 010 101 010 101 010 101 0
2	−0.001 656 565 656 565 656 565 656 565	0.000 001 182 026 143 790 849 673 202 6
3	−0.001 657 747 682 709 447 415 329 768	0.000 000 714 819 415 688 980 906 372 2
4	−0.001 657 032 863 293 758 434 423 396	0.000 001 375 572 238 237 142 451 701 8
5	−0.001 658 408 435 531 995 576 875 097	0.000 006 482 975 684 933 996 045 107 1
6	−0.001 651 925 459 847 061 580 829 990	0.000 063 464 368 855 638 413 105 027 2
7	−0.001 715 389 828 702 699 993 935 017	0.001 152 174 438 923 418 820 406 725 6
8	−0.000 563 215 389 779 281 173 528 292	0.035 707 817 029 811 724 407 831 959 7

TABLE 4. Numerical values for the algebraic series and its remainder with $a = 0.2$

N	$\mathcal{T}_3^L(0.4, N)$	R_N
1	−0.003 333 333 333 333 333 333 333 333	0.000 080 808 080 808 080 808 080 808 0
2	−0.003 252 525 252 525 252 525 252 525	0.000 037 824 836 601 307 189 542 483 6
3	−0.003 290 350 089 126 559 714 795 008	0.000 091 496 885 208 189 556 015 642 9
4	−0.003 198 853 203 918 370 158 779 365	0.000 704 292 985 977 416 935 271 341 4
5	−0.003 903 146 189 895 787 094 050 707	0.013 277 134 202 744 823 900 379 455 9
6	0.009 373 988 012 849 036 806 328 748	0.519 900 109 665 389 880 156 382 933 7
7	−0.510 526 121 652 540 843 350 054 185	37.754 452 014 642 587 907 087 587 540

TABLE 5. Numerical values for the algebraic series and its remainder with $a = 0.4$

a	N_L	$\mathcal{T}_3^L(a, N_L)$
0.01	15	$-8.333\,207\,107\,589\,808\,008\,820\,189\,298\,0 \times 10^{-5}$
0.2	2	$-1.656\,565\,656\,565\,656\,565\,656\,565\,656\,6 \times 10^{-3}$
0.4	1	$-3.333\,333\,333\,333\,333\,333\,333\,333\,333\,3 \times 10^{-3}$
1	1	$-8.333\,333\,333\,333\,333\,333\,333\,333\,333\,3 \times 10^{-3}$
10	1	$-8.333\,333\,333\,333\,333\,333\,333\,333\,333\,3 \times 10^{-2}$

TABLE 6. The algebraic series terminated at its optimal number of terms N_L

a	N_L	$\mathcal{T}_3(a) - \mathcal{T}_3^L(a, N_L)$
0.01	15	$7.804\,439\,506\,214\,946\,082\,138\,031\,101\,7 \times 10^{-20}$
0.2	2	$-1.703\,072\,002\,458\,385\,547\,892\,707\,803\,9 \times 10^{-4}$
0.4	1	$2.476\,324\,127\,429\,265\,130\,307\,072\,093\,1 \times 10^{-3}$
1	1	$-1.643\,127\,443\,469\,151\,563\,434\,673\,758\,3 \times 10^{-2}$
10	1	$1.688\,943\,603\,086\,137\,740\,818\,247\,315\,4 \times 10^{-1}$

TABLE 7. The remainder after subtraction of the algebraic series terminated at its optimal number from the values for the tail of $S_3(a)$

n	$\Delta T_3^L(0.01, 15, n)$	$\Delta T_3^L(0.01, 15, n)_E$
1	$1.271\,689\,3 \times 10^{-27}$	$1.259\,586\,851\,355\,197\,244\,866\,683\,869\,854\,038\,555\,4 \times 10^{-27}$
2	$1.315\,987\,8 \times 10^{-55}$	$1.303\,846\,026\,007\,3 \times 10^{-55}$
3	$3.751\,743\,3 \times 10^{-72}$	$3.717\,165\,228\,083\,9 \times 10^{-72}$

TABLE 8. The first three terms of the terminant sum for the algebraic series with $a = 0.01$

n	$\Delta T_3^L(0.2, 2, n)$	$\Delta T_3^L(0.2, 2, n)_E$
1	$-9.121\,784\,7 \times 10^{-7}$	$-8.458\,291\,175\,043\,039\,637\,227\,193\,577\,874\,183\,5 \times 10^{-7}$
2	$-1.915\,216\,9 \times 10^{-11}$	$-1.787\,037\,897\,563\,896\,317 \times 10^{-11}$
3	$-2.939\,838\,0 \times 10^{-14}$	$-2.743\,600\,619\,993\,147\,32 \times 10^{-14}$

TABLE 9. The first three terms of the terminant sum for the algebraic series with $a = 0.2$

n	$\Delta T_3^L(0.4, 1, n)$	$\Delta T_3^L(0.4, 1, n)_E$
1	$7.352\,968\,6 \times 10^{-5}$	$6.325\,502\,923\,986\,384\,224 \times 10^{-5}$
2	$8.974\,132\,1 \times 10^{-8}$	$7.827\,829\,149\,258\,315\,55 \times 10^{-8}$
3	$1.565\,679\,6 \times 10^{-9}$	$1.366\,256\,434\,842\,610\,44 \times 10^{-9}$

TABLE 10. The first three terms of the terminant sum for the algebraic series with $a = 0.4$

n	$\Delta T_3^L(1, 1, n)$	$\Delta T_3^L(1, 1, n)_E$
1	$7.450\,864\,1 \times 10^{-4}$	$6.252\,805\,360\,209\,226\,847 \times 10^{-4}$
2	$1.360\,347\,8 \times 10^{-6}$	$1.184\,188\,625\,579\,520\,531 \times 10^{-6}$
3	$2.438\,822\,1 \times 10^{-8}$	$2.127\,730\,945\,764\,719\,76 \times 10^{-8}$

TABLE 11. The first three terms of the terminant sum for the algebraic series with $a = 1$

n	$\Delta T_3^L(10,1,n)$	$\Delta T_3^L(10,1,n)_E$
1	$5.973\,407\,3 \times 10^{-2}$	$4.303\,335\,159\,832\,884\,52 \times 10^{-2}$
2	$6.225\,574\,4 \times 10^{-4}$	$5.178\,795\,131\,570\,912\,41 \times 10^{-4}$
3	$1.990\,454\,6 \times 10^{-5}$	$1.714\,628\,191\,929 \times 10^{-5}$

TABLE 12. The first three terms of the terminant sum for the algebraic series with $a = 10$

N_K	$T_3^K(0.01, N_K)$
1	$7.914\,855\,607\,745\,877\,772\,547\,326\,276\,512\,530\,9 \times 10^{-20}$
2	$7.803\,140\,505\,323\,091\,064\,820\,048\,153\,055\,235\,1 \times 10^{-20}$
3	$7.804\,454\,123\,887\,647\,578\,225\,741\,395\,506\,214\,6 \times 10^{-20}$
4	$7.804\,439\,400\,962\,753\,479\,176\,795\,680\,214\,931\,9 \times 10^{-20}$
5	$7.804\,439\,367\,174\,593\,282\,755\,066\,501\,436\,133\,2 \times 10^{-20}$
6	$7.804\,439\,381\,004\,887\,979\,031\,623\,309\,297\,701\,1 \times 10^{-20}$
7	$7.804\,439\,380\,241\,040\,619\,201\,846\,648\,551\,482\,1 \times 10^{-20}$
8	$7.804\,439\,380\,266\,225\,965\,273\,372\,598\,242\,055\,0 \times 10^{-20}$
9	$7.804\,439\,380\,266\,359\,616\,593\,161\,132\,668\,357\,7 \times 10^{-20}$
10	$7.804\,439\,380\,266\,250\,748\,258\,602\,119\,513\,728\,3 \times 10^{-20}$
11	$7.804\,439\,380\,266\,261\,538\,047\,313\,104\,077\,480\,9 \times 10^{-20}$
12	$7.804\,439\,380\,266\,260\,946\,325\,952\,763\,551\,378\,2 \times 10^{-20}$
13	$7.804\,439\,380\,266\,260\,941\,398\,548\,956\,811\,750\,0 \times 10^{-20}$
14	$7.804\,439\,380\,266\,260\,947\,412\,879\,252\,018\,903\,2 \times 10^{-20}$
15	$7.804\,439\,380\,266\,260\,946\,552\,557\,144\,724\,221\,0 \times 10^{-20}$
16	$7.804\,439\,380\,266\,260\,946\,618\,570\,613\,871\,156\,2 \times 10^{-20}$
17	$7.804\,439\,380\,266\,260\,946\,619\,319\,863\,564\,236\,4 \times 10^{-20}$
18	$7.804\,439\,380\,266\,260\,946\,618\,100\,959\,155\,930\,5 \times 10^{-20}$
19	$7.804\,439\,380\,266\,260\,946\,618\,328\,896\,143\,814\,1 \times 10^{-20}$
20	$7.804\,439\,380\,266\,260\,946\,618\,306\,414\,920\,190\,9 \times 10^{-20}$
21	$7.804\,439\,380\,266\,260\,946\,618\,306\,091\,785\,374\,4 \times 10^{-20}$

TABLE 13. The K-part of the tail of $S_3(0.01)$ for various values of N_K

N_K	$\mathcal{T}_3^K(0.2, N_K)$	$\mathcal{T}_3^K(0.4, N_K)$
1	$-1.702\,147\,279\,865\,673\,402\,443\,3 \times 10^{-4}$	$2.424\,174\,516\,210\,200\,978 \times 10^{-3}$
2	$-1.694\,520\,596\,627\,128\,945\,835\,0 \times 10^{-4}$	$2.413\,022\,225\,567\,857\,318 \times 10^{-3}$
3	$-1.694\,602\,074\,671\,519\,163\,122\,7 \times 10^{-4}$	$2.412\,922\,262\,629\,495\,199 \times 10^{-3}$
4	$-1.694\,616\,086\,164\,045\,714\,168\,6 \times 10^{-4}$	$2.413\,003\,766\,370\,671\,286 \times 10^{-3}$
5	$-1.694\,613\,179\,608\,181\,474\,193\,2 \times 10^{-4}$	$2.412\,987\,208\,521\,719\,171 \times 10^{-3}$
6	$-1.694\,613\,557\,280\,577\,846\,833\,7 \times 10^{-4}$	$2.412\,989\,417\,558\,648\,662 \times 10^{-3}$
7	$-1.694\,613\,538\,329\,326\,673\,372\,5 \times 10^{-4}$	$2.412\,989\,510\,561\,536\,403 \times 10^{-3}$
8	$-1.694\,613\,528\,741\,984\,722\,838\,3 \times 10^{-4}$	$2.412\,989\,287\,486\,333\,132 \times 10^{-3}$
9	$-1.694\,613\,533\,340\,814\,905\,183\,1 \times 10^{-4}$	$2.412\,989\,392\,279\,407\,788 \times 10^{-3}$
10	$-1.694\,613\,532\,151\,640\,973\,576\,5 \times 10^{-4}$	$2.412\,989\,364\,457\,103\,629 \times 10^{-3}$
11	$-1.694\,613\,532\,258\,719\,956\,466\,9 \times 10^{-4}$	$2.412\,989\,362\,355\,151\,714 \times 10^{-3}$
12	$-1.694\,613\,532\,348\,820\,525\,345\,6 \times 10^{-4}$	$2.412\,989\,370\,740\,876\,467 \times 10^{-3}$
13	$-1.694\,613\,532\,281\,001\,387\,059\,8 \times 10^{-4}$	$2.412\,989\,364\,559\,325\,304 \times 10^{-3}$
14	$-1.694\,613\,532\,307\,279\,311\,975\,6 \times 10^{-4}$	$2.412\,989\,367\,018\,553\,176 \times 10^{-3}$
15	$-1.694\,613\,532\,303\,864\,141\,626\,1 \times 10^{-4}$	$2.412\,989\,367\,286\,711\,265 \times 10^{-3}$
16	$-1.694\,613\,532\,299\,843\,430\,850\,8 \times 10^{-4}$	$2.412\,989\,365\,789\,869\,490 \times 10^{-3}$
17	$-1.694\,613\,532\,303\,968\,399\,742\,2 \times 10^{-4}$	$2.412\,989\,367\,293\,793\,420 \times 10^{-3}$
18	$-1.694\,613\,532\,301\,838\,135\,726\,2 \times 10^{-4}$	$2.412\,989\,366\,496\,347\,707 \times 10^{-3}$
19	$-1.694\,613\,532\,302\,200\,066\,977\,7 \times 10^{-4}$	$2.412\,989\,366\,382\,672\,800 \times 10^{-3}$
20	$-1.694\,613\,532\,302\,747\,776\,402\,7 \times 10^{-4}$	$2.412\,989\,367\,198\,284\,162 \times 10^{-3}$
21	$-1.694\,613\,532\,302\,036\,173\,273\,3 \times 10^{-4}$	$2.412\,989\,366\,160\,509\,572 \times 10^{-3}$

TABLE 14. The K-part of the tail of $S_3(0.2)$ and $S_3(0.4)$ for various values of N_K

N_K	$\mathcal{T}_3^K(1, N_K)$	$\mathcal{T}_3^K(10, N_K)$
1	$-1.707\,946\,071\,845\,369 \times 10^{-2}$	$1.248\,570\,961\,57 \times 10^{-1}$
2	$-1.707\,059\,786\,107\,426 \times 10^{-2}$	$1.265\,362\,404\,61 \times 10^{-1}$
3	$-1.705\,444\,284\,878\,370 \times 10^{-2}$	$1.244\,968\,746\,91 \times 10^{-1}$
4	$-1.705\,843\,114\,645\,748 \times 10^{-2}$	$1.258\,026\,182\,46 \times 10^{-1}$
5	$-1.705\,770\,181\,303\,465 \times 10^{-2}$	$1.252\,022\,593\,70 \times 10^{-1}$
6	$-1.705\,771\,282\,350\,021 \times 10^{-2}$	$1.250\,068\,342\,64 \times 10^{-1}$
7	$-1.705\,780\,675\,804\,986 \times 10^{-2}$	$1.261\,963\,559\,52 \times 10^{-1}$
8	$-1.705\,773\,853\,343\,695 \times 10^{-2}$	$1.239\,565\,868\,51 \times 10^{-1}$
9	$-1.705\,776\,738\,284\,197 \times 10^{-2}$	$1.263\,356\,495\,56 \times 10^{-1}$
10	$-1.705\,776\,651\,608\,086 \times 10^{-2}$	$1.278\,724\,263\,56 \times 10^{-1}$
11	$-1.705\,775\,324\,728\,797 \times 10^{-2}$	$1.110\,689\,213\,84 \times 10^{-1}$
12	$-1.705\,776\,927\,643\,452 \times 10^{-2}$	$1.636\,938\,196\,65 \times 10^{-1}$
13	$-1.705\,775\,864\,034\,961 \times 10^{-2}$	$7.598\,099\,416\,02 \times 10^{-2}$
14	$-1.705\,775\,911\,918\,416 \times 10^{-2}$	$-8.915\,470\,964\,87 \times 10^{-3}$
15	$-1.705\,776\,969\,903\,521 \times 10^{-2}$	$1.330\,910\,378\,223 \times 10^{0}$
16	$-1.705\,775\,181\,663\,917 \times 10^{-2}$	$-4.540\,019\,200\,722 \times 10^{0}$
17	$-1.705\,776\,798\,962\,534 \times 10^{-2}$	$8.797\,397\,272\,162 \times 10^{0}$
18	$-1.705\,776\,701\,918\,719 \times 10^{-2}$	$2.600\,307\,851\,684 \times 10^{1}$
19	$-1.705\,773\,898\,853\,032 \times 10^{-2}$	$-3.289\,754\,456\,886 \times 10^{2}$
20	$-1.705\,779\,988\,794\,213 \times 10^{-2}$	$1.670\,399\,588\,433 \times 10^{3}$
21	$-1.705\,773\,013\,743\,083 \times 10^{-2}$	$-4.081\,733\,103\,504 \times 10^{3}$

TABLE 15. The K-part of the tail of $S_3(1)$ and $S_3(10)$ for various values of N_K

a	N_L	N_K	$\mathcal{T}_3(a) - \mathcal{T}_3^L(a, N_L) - \mathcal{T}_3^K(a, N_K)$
0.01	15	21	$1.259\,586\,851\,355\,197\,250\,099\,309\,733 \times 10^{-27}$
0.2	2	19	$-8.458\,470\,156\,185\,480\,915 \times 10^{-7}$
0.4	1	18	$6.333\,476\,093\,291\,742\,3 \times 10^{-5}$
1	1	3	$6.231\,684\,140\,921\,9 \times 10^{-4}$
10	1	13	$9.291\,336\,614\,83 \times 10^{-2}$

TABLE 16. The remainder $\mathcal{R}_3(a, N_L, N_K)$ after both the algebraic and exponential series have been removed from the tail for $S_3(a)$

a	$\Delta\mathcal{T}_3^L(a,N_L)_E$	$\mathcal{R}_3(a,N_L,N_K)-\Delta\mathcal{T}_3^L(a,N_L)_E$
0.01	$1.259\,586\,851\,355\,197\,244\,866\,683\,86\times10^{-27}$	$5.232\,625\,863\,18\times10^{-45}$
0.2	$-8.458\,470\,156\,026\,585\,381\,5\times10^{-7}$	$-1.588\,955\,334\times10^{-17}$
0.4	$6.333\,476\,076\,615\,190\,6\times10^{-5}$	$1.667\,655\,16\times10^{-13}$
1	$6.264\,873\,602\,519\,64\times10^{-4}$	$-3.318\,946\,159\,774\times10^{-6}$
10	$4.356\,966\,629\,2\times10^{-2}$	$4.934\,369\,985\,6\times10^{-4}$

TABLE 17. Removal of the terminant sum for the algebraic series from $\mathcal{R}_3(a,N_L,N_K)$

a	$\Delta\mathcal{T}_3^K(a,N_K)_E$
0.01	$5.232\,625\,863\,18\times10^{-45}$
0.2	$-1.588\,955\,334\times10^{-17}$
0.4	$1.667\,655\,16\times10^{-13}$
1	$-3.318\,946\,159\,774\times10^{-6}$
10	$4.934\,369\,985\,6\times10^{-4}$

TABLE 18. Values of the terminant sum for the exponential series in $S_3(a)$

	formula	integral cut-off	precision	accuracy	CPU time
$S_3(a)$			9 digits	10^{-12}	0.73 s
$\mathcal{T}_3^L(a)$	(7.7)		10 digits	10^{-12}	0.08 s
$\Delta\mathcal{T}_3^L(a,N_L)_E$	(8.14)	$\int_{-1/4}^{-1/4+3i} dt$	8 digits	10^{-12}	1.42 s
$\Delta\mathcal{T}_3^L(a,N_L)_E$	(7.18)	$\int_0^1 ds \int_0^{50} dt \int_0^{50} dy$	8 digits	10^{-12}	60712 s
$\mathcal{T}_3^K(a)$	(7.20)		10 digits	10^{-12}	0.31 s
$\Delta\mathcal{T}_3^K(a,N_K)_E$	(8.15)	$\int_{-1/10-6i}^{-1/10+6i} dt$	8 digits	10^{-12}	29.27 s
$\Delta\mathcal{T}_3^K(a,N_K)_E$	(7.26)	$\int_0^1 dt \int_0^{20} dy$	8 digits	10^{-12}	66.67 s

TABLE 19. CPU times for the computation of $S_3(a)$, its dominant asymptotic expansion $\mathcal{T}_3^L(a)$, the corresponding terminant sum $\Delta\mathcal{T}_3^L(a,N_L)_E$, its subdominant expansion $\mathcal{T}_3^K(a)$, and the corresponding terminant sum $\Delta\mathcal{T}_3^K(a,N_K)_E$, with $a=0.4$, $N_L=1$ and $N_K=1$.

k	C_1	C_2	γ_2
3	$2^{-1/4}3^{-1/4}$	0	0
4	$2^{-1/6}3^{-1/2}$	0	0
5	$2^{-7/8}5^{-1/8}$	0	0
6	$3^{-1/10}5^{-1/2}$	$2^{-1}3^{-1/10}5^{-1/2}$	1
7	$2^{-5/12}3^{-1/2}7^{-1/12}$	$2^{-5/12}3^{-1/2}7^{-1/12}$	11/12

TABLE 20. Parameters C_1, C_2, and γ_2 for the exact terminants $\Delta\mathcal{T}_k^K(a, N_K)_E$ given in Eq. (8.19).

Index

Printed in the United States
By Bookmasters